THINKING THROUGH DATA

• • •   **Sensing Media**
Aesthetics, Philosophy,
and Cultures of Media
EDITED BY WENDY HUI KYONG CHUN
AND SHANE DENSON

THINKING THROUGH DATA

How Outliers, Aggregates, and Patterns Shape Perception

MAJA BAK HERRIE

STANFORD UNIVERSITY PRESS

Stanford, California

Stanford University Press
Stanford, California

Library of Congress Cataloging-in-Publication Data

Names: Bak Herrie, Maja, author.

Title: Thinking through data : how outliers, aggregates, and patterns shape perception / Maja Bak Herrie.

Other titles: Sensing media (Series)

Description: Stanford, California : Stanford University Press, 2025. | Series: Sensing media | Includes bibliographical references and index. |

Identifiers: LCCN 2024040294 (print) | LCCN 2024040295 (ebook) | ISBN 9781503641891 (cloth) | ISBN 9781503642331 (paperback) | ISBN 9781503642102 (ebook)

Subjects: LCSH: Statistics and aesthetics. | Aesthetics, Modern—21st century. | Art and technology.

Classification: LCC HA30.3 .B37 2025 (print) | LCC HA30.3 (ebook) | DDC 330.01/5195—dc23/eng/20241115

LC record available at https://lccn.loc.gov/2024040294

LC ebook record available at https://lccn.loc.gov/2024040295

Cover design: Bob Aufuldish / Aufuldish & Warinner

Cover photographs: Mikkel Andreas Junker Jensen, 2015, Glamsbjerg, Denmark

To Elise & Ebbe

CONTENTS

ACKNOWLEDGMENTS

Cultivating quality research requires more than individual effort. Rather, it flourishes within nurturing and collaborative settings, within and beyond the academy. Therefore, I am deeply grateful to the colleagues, friends, and family members whose unwavering support has made this book a reality. I have been fortunate to be part of several enriching research groups and programs, and therefore, I would like to thank all the wonderful people in Centre for Aesthetics of AI Images (AIIM), Digital Aesthetics Research Centre (DARC), in the research programs Contemporary Aesthetics and Technology (CAT) and Arts, Aesthetics and Communities (ACC), and at the Department for Art History, Aesthetics and Culture, and Museology.

A special thanks to all of you who generously read, thought, and sparred with me along the way: Aurora Hoel, Elisabeth Brun, Tobias Dias, Asker Bryld Staunæs, Sophus Helle, Lotte Philipsen, Anne Kølbæk Iversen, Thomas Erslev, Maj Ørskov, Lea Laura Nørregaard Michelsen, Lone Koefoed Hansen, Søren Pold, Andreas Ervik, Nanna Bonde Thylstrup, Mikkel Thorup, Nicolas Malevé, Mette-Marie Zacher Sørensen, Birgitte Stougaard Pedersen, Jacob Lund, Liv Hausken, Frederik Tygstrup, Sarah Yazdani, and Steffen Krüger. Also, a warm thanks to my dear family and friends for your patience and outstanding listening skills.

Finally, a sincere thanks to you, Jørn Erslev Andersen, for so attentively and competently guiding me through the first years as a researcher, for being an inexhaustible source of inspiration and enthusiasm, and for being one of the most caring people I know. And finally, thanks, Simon Enni, both for lending me your knowledge-power within the convolutions of data science and mathematics, and for your endless patience, attention, and support. We think and understand together, and you are present on every page of this book.

Introduction

THINKING THROUGH DATA

••• A politician brings a chart to Parliament, which shows that the economy is growing slower than expected—we need to cut costs. On another screen, another graph shows the kind of growth that is cause for concern: global CO_2 emissions are not slowing down—we need to act now. Statistics show that the number of refugees from North Africa and the Middle East is in the millions—where can they go? Data and numerical representations are ubiquitous in modern society, and we meet them in a range of platforms and in many contexts: in financial sector predictions, in political debates, in insurance and risk management, in immigration policies, biometrics, medicine, and more. Statistical entities help us make sense of connections that would otherwise seem random and see patterns that would otherwise be invisible. Yet, we often forget to ask: What does it mean to see the world through a curve? How do data shape our thinking about the world around us?

Statistical entities such as averages, aggregates, distributions, outliers, and patterns seem to hover between several domains: they are neither invented nor discovered, they are neither real nor metaphysical. On the one hand, they are "creatures of classification and calculation, of conventions of coding, modeling, and sampling. It is the artifice of definition that makes them cohere—or unravel," as science historian

Lorraine Daston states.[1] They acquire significance and strength in specific historical circumstances and pass away in others. On the other hand, statistical entities are also robust. The search for averages, as the statistician Adolphe Quetelet led from around the 1830s and onwards, was not only influential for the natural sciences, but also marked the beginning of an extensive interest in regularities in statistics—in demography, criminology, and medicine—from the perspective of large populations and the law of large numbers.[2] Mathematician and physicist Carl Friedrich Gauss's famous distribution, the one sometimes referred to as "the normal distribution," would turn out to be not just one of the most important and frequently occurring probability distributions in statistics, but also a completely novel way of approaching the world through data, where measurements from different contexts distribute in the same way, as a bell curve around an estimated mean value.[3] The search for orderly shapes in a chaotic world had begun.

VISUAL REASONING

Turning to statistic's curves, lines, graphs, and bells, this book addresses the aesthetic dimensions of statistical and data-driven knowledge production. It explores the background of today's so-called data-driven paradigm[4] and explains how elusive yet crucial statistical ideas, such as outliers, aggregates, and patterns, form how we perceive and make sense of data. From the sixteenth century's measurements of the foot, to the blurred facial features of *L'Homme Moyen*, to the image aggregates of today's machine vision output, the examples collected in this book illustrate the central role of aesthetics throughout the history of statistical knowledge production. However artificial or calculated, statistical entities change the world. This book sets out to complicate and challenge purely technical conceptions of statistical data processing by asking, in various ways and from various perspectives, how knowledge is produced, based on statistics and data, and how it manifests as knowledge. What are we thinking *with* when we use digital calculations and models, and how is digital knowledge production to become possible and legitimate?

The foregoing questions are addressed through a discussion of the status of knowledge production as a fundamentally mediated and aesthetic phenomenon. Knowledge is mediated, as mathematical and statistical thinking rely on abstract models and symbolic systems that decisively affect and shape analytical outcomes. When the bell shape of a curve frames analyses and interpretations of complex phenomena such as economics or demographics, most measurements are expected to distribute around a mean, with rare exceptions on the tail ends to either side. This expectation ultimately affects what is gathered, what is measured, and what is sought: it includes those aspects that conform to the bell shape, while excluding those that do not. As such, this modeled view of the measured input should be understood as more than a mere vehicle of information transfer, along the lines of classical theories of communication. As media theorist Liv Hauksen argues, it is necessary to move beyond the idea of a medium as something rather fixed, like an apparatus, and toward conceptions of mediation as a process, as the performance of a task. Such a "shift from medium to mediation does not only represent a shift in focus," she argues, but also "represents a shift in perspective from medium as an object of study and media as collections of artifacts and technologies, to medium and media as concepts, ideas, models for understanding practices, articulations and experiences."[5] By emphasizing such a focus on mediation, this book explores the contexts, discourses, and cultural systems that situate the knowledge production of statistical data processing. It emphasizes that today's so-called data-driven and empiricist epistemologies cannot be reduced to, or deduced from, the techniques used and the technologies of which they are a part.[6] In addition to this book's focus on processes of mediation, it presents a discussion of the status of knowledge production as a fundamentally aesthetic phenomenon. By applying the fundamental instantiations of these abstractions as visualizations and diagrams, it investigates the concrete processes by which knowledge is produced and made generalizable.[7] In this book, aesthetics is understood as broader than just something pertaining to the realm of art and art criticism.[8] Instead, it is linked to knowledge production and science more generally and describes a certain way of knowing, a particular attitude,[9] or disponibility for a critical, aesthetic way of obtaining knowledge.[10] Put differently, this book studies a range of concrete phenomena—graphs, scientific diagrams, mathematical visualizations, and artistic investigations and artworks that address such mathematical and statistical entities—by examining their particular ways of operating aesthetically and uses these insights to investigate the more general mode of operation by which knowledge comes into being. Thus, mathematical and statistical entities present certain insights through complex processes of mediation, and in doing so, they allow scholars to approach the subjects they study. Crystallizing in a particular "line of sight," as A. S. Aurora Hoel poetically expresses it,[11] a graph, diagram, or visualization does not only transmit information, but also expressively conveys meaning, making it visible, and thereby understandable.[12]

FORMAL OBJECTS OF THOUGHT

By examining statistical entities and the general ideas or ways of thinking to which they relate, this book proposes a theoretical concept, the digital object, which runs throughout its entirety: digital, because it describes a particular digital (numerical, computational) mode of operation; object, because this mode of operation always assumes a tentative form (grammatically, phenomenologically, dispositively), so to speak, because it may be seen as the "thing" to which events, actions, and experiences adjust. Viewed in this light, the digital object may be said to approach its subjects in a similar way to Foucault's apparatus (*dispositif*).[13] It engenders a new, distinctive way of knowing, grounded in a sensory depiction of the world—new aesthetic, social, and political behavioral patterns attach themselves to data-driven knowledge development. As such, we are dealing with a collective "set of strategies of the relations of forces supporting, and supported by" certain types of data-driven knowledge,[14] whether this is achieved through standardization strategies designed to moderate and sort demographic distributions, aggregated facial data-mapping utilized to surveil and control, or comparisons meant to twist and turn the data set into appropriate patterns.

The digital object suggested here is not an object in the usual sense, but a theoretical concept set up to investigate a heterogeneous digital field that could not be coherently described in other ways. It may be understood as a conceptual intervention: a (con)*cept* (from the Greek *kope*, meaning "oar" or "handle," or the Latin *capere*, meaning "to grasp, lay hold, or comprehend"),[15] that is, something to hold on to or grasp. It is analogous to Karl Marx's conception of political economy as a theoretical object expressed through the interactions of commodities, production, and labor. In this sense, a digital object is not a concrete object such as a compiler, a desktop icon, or a fiberoptic cable, but the thing that digital entities such as outliers, aggregates, and patterns are expressions of, in the same way that for Marx, commodities are expressions of political economy.[16] Similarly to other theoretical objects such as *la langue* in Ferdinand de Saussure's framework,[17] which do not exist "out there" in the physical, concrete sense—and seem to elude direct measurement and empirical observation—a digital object also needs to be applied analytically for the purpose of seeing something that would otherwise remain obscure, or to assemble something that would remain fragmented. Just as Marx's famous critique of political economy identifies important political questions, such as exploitation and alienation, by analyzing the relations between commodities, production, and labor, in this book, analyses of a digital object's manifestations as outliers, aggregates, and patterns are used to reveal important epistemological questions. Although it is difficult, if not impossible, to study abstract-formal modes of operation directly—political economy as such or the fundamental nature of language—they may be approached by analyzing the ways in which they are realized concretely. Therefore, each of the three main chapters of this book is dedicated to a specific mode of the digital object's operation—the outlier, the aggregate, and the pattern, respectively—with the purpose of understanding how the digital object influences and shapes modern knowledge production.

The definition of the digital object presented in this book diverges from previous definitions of digital objects by referencing a theoretical concept, rather than a list of things "out there." It is not a collection or compilation of a range of machines or apparatuses creating meaning through their physical components or material composition as in, for example, Jussi Parikka's material studies of digital "machines" or Wolfgang Ernst's media archaeological works.[18] It does not describe a selection of semi-physical formations, composed of data and metadata structured within a computational assemblage, which assume the semblance of objects, as theorized by Yuk Hui in his exploration of digital objects, influenced by the work of Gilbert Simondon and Bernard Stiegler.[19] And it does not equate (big) data to big objects akin to Timothy Morton's hyperobjects such as global warming, climate, or oil. Instead, the digital object presented in this book transcends ontological delineations, existing foremost as an epistemological construct—a conceptual tool employed to reveal a range of processes or structural connections that take on the form of an object that permeates modern processes of knowledge production in analytical work, theoretical traditions, and concrete applications. It serves as a unit of understanding, facilitating both theoretical analysis and practical application—the ability to transform an experience of the world into an unambiguous digital representation, and to share these digital data with various communities of knowledge, or apply them to politics or economics. The digital object changes our way of seeing the world, whether by transforming the dimensions of a house into a number of standardized feet, the people of a country into a distribution of demographic data, or a person's face into a robust biometric model.

CULTURAL TECHNIQUES OF PERCEPTION

In some ways, the concept of the digital object, as theorized about and discussed in this book, is akin to theories of cultural techniques (*Kulturtechniken*)[20] linked to developments in Germanophone media theory since the 1970s. It originated in the realm of agriculture and has been used to describe large-scale soil improvements (*cultivation*),[21] such as irrigation and drainage, straightening riverbeds, or constructing water reservoirs.[22] Today, however, the term is associated with cultural and medial questions, and describes interactions between humans and media, although still with a general focus on their technical aspects. As such, cultural techniques may be understood as operative procedures that transcend medial boundaries, or, as the philosopher Sybille Krämer and the art historian Horst Bredekamp have stated:

> [C]ultural techniques are (a) operative processes that enable work with things and symbols; (b) they are based on a separation be-

tween an implied "know how" and an explicit "know that"; (c) they can be understood as skills that habituate and regularize the body's movements and that express themselves in everyday fluid practices; (d) at the same time, such techniques can provide the aesthetic and material-technical foundation for scientific innovation and new theoretical objects.[23]

Having read Krämer and Bredekamp, one might argue that cultural techniques are situated between process and thing, that is, between established and regulated patterns, shaped and sedimented over time, and more established thing-like formations that, via scientific processes, for example, acquire an aesthetic, or sometimes material, character. This technical-structural interest in cultural phenomena inherent in theories of cultural techniques—particularly in Krämer and Bredekamp's work—aligns with this book's aim of developing a digital version of the better-known theoretical object encountered in theories such as Ferdinand de Saussure's abstract language system or Karl Marx's political economy. Krämer and Bredekamp state that this is an approach that seeks an understanding of the fundamental "physiognomy of a culture" by investigating the structural operations that regulate and organize things: "The history of culture always already is the history of its cultural techniques," they write, "just as the history of science cannot be decoupled from the changes in the everyday techniques of perception, communication, representation, archiving, counting, measuring."[24] Cultural techniques define the agency of media, or, in media theorist Cornelia Vismann's words, "If media theory were, or had, a grammar, that agency would find its expression in objects claiming the grammatical subject position and cultural techniques standing in for verbs."[25]

The persistent interest in the technologically mediated grammar of culture is particularly interesting in the context of this book, because it supplements a (historically Anglo-American) focus on the content or meaning of media or things by suggesting that we take a closer look at what media theorist Liam Cole Young calls technical-scientific "ways of doing": counting, measuring, collecting, observing, playing, confessing, listing.[26] What we call media (gramophones, telegraphs, and computers, to use Young's examples) communicate and order things around them by encoding non-sense into sense (and vice versa). For instance, by translating data into alphanumerical characters, a computer does not only transmit, it also transforms. Cultural techniques—and the concept of the digital object proposed here—assume the position of the parasitic third,[27] as media theorist Bernhard Siegert—via philosopher and mathematician Michel Serres—states; that is, a kind of middle or mediating position that operates through a communicative work that disassociates itself from any dichotomic form. In Serres's model of communication, the fundamental relationship is not between sender and receiver—as it so often is in classical communication models—but between communication and *noise*.[28] For Siegert (and Serres), communication is not primarily information exchange, but an act of ordering that introduces distinctions and differentiations. Siegert's understanding of cultural techniques is one of interruption, disturbance, and deviation,[29] that is, a way of approaching media that aims to create "an awareness of the plenitude of a world of as-yet-undistinguished things that, as an inexhaustible reservoir of possibilities, remains the basic point of reference for every type of culture," as he writes in his 2015 book, *Cultural Techniques: Grids, Filters, Doors, and Other Articulations of the Real*.[30]

Although this book is deeply inspired by theories of cultural techniques, it also differs from them in decisive ways. An example of a technical mode of operation that displays this difference could be listing. As an artifact, the list is found as far back on Ancient Sumerian clay tablets, but it is also present in contemporary computer code. If one addresses listing as a cultural technique, one becomes aware of how it not only distributes data, but also defines certain items, inscribes order, and helps to decide what to include and exclude.[31] Listing transforms people, words, or things into dynamic units that may be processed, stored, or transmitted. Listing determines which words are significant or redundant when included in searches, for example, when using Google's search engine.[32] Listing encodes things into a symbolic order and thereby subjects them to manipulation, revision, erasure, or reversibility.[33] In many cases, listing is more political[34] than one might think: as Young states, this form of protocol "determines how computation unfolds; how a person is listed can determine his or her fate."[35] However, listing, as a technical mode of operation, or a cultural technique in the digital age, also implies a theoretical object that—so to speak—"takes" this technique as its thing. What kind of object enables this particular alphanumerical encoding of the symbolic order? What kind of object translates people, words, or things into units that may be distributed, transformed, or deleted? What kind of thinking goes into a protocol or a list? Listing, as phenomenon, may be studied in several ways: it may be examined through the lens of media archaeology—as a concrete machine or technique that, through material appliances and historical sediments, provides a range of varied, overlooked potential;[36] it may be investigated in light of systemic or computational operations, such as arrays, queues, and stacks, and in terms the databases that organize it;[37] it may explored in terms of its poetic relation to language, literature, and imagination, as Umberto Eco[38] and Jorge Luis Borges[39] did; or it may, as mentioned above, be studied as a cultural technique that has a certain kind of medial agency in its fundamental communicative potential. The list as the concrete thing that manipulates, revises, erases, or reverses is decisive for the manifestation of the listing *logic* (the theoretical object) behind it, which produces its epistemological, symbolic, or political effects. Here, again, it is not so much about the physical configuration of the list—the clay tablet or the code fragments—as it is about the list as grammatical subject that acts on behalf of the more general listing operation, that is, the necessary, yet ephemeral basis for a digital way of thinking, or the appearance of a list as a concrete, aesthetic phenomenon that contributes to knowledge production. This epistemological object is the recurring figure in this book. Conceptually, it should not be understood as a comprehensive categorization, or a typological designation that seeks to conform or sort a range of phenomena. As a contribution, its form is way more abductive; more of an experimental, theoretical, and analytical synthesis than an exhaustive description of an emerging digital field.

COUNTING ON YOUR FINGERS

The term *digital* is central to this book and therefore requires a separate, brief introduction. The word *digital* is primarily used to describe things and experiences that have been digitalized, that is, where information has been extracted and encoded as digital data.[40] This new numerical representation has one great feature: the ability to collect, compile, and analyze data sets, to describe them statistically (descriptive analytics), to mine them for knowledge (explorative analytics), to test a hypothesis (inferential analytics), to use them for prediction (predictive analytics), or to employ them to control or govern (prescriptive analytics). Recent years have also seen a surge in interest in generative practices, where identified statistical patterns are used to expand a data set with new, simulated members, whether they be images, texts, or something else entirely. These uses of digital data indicate that an extractive logic is at work, one that abstracts and formalizes experiences of the world for the purpose of processing it. However, the etymological root of the word *digital* suggests a different interpretation of digital data—one that is inescapably tied to the context and situation of the digitized experience.

Etymologically, the word *digital* comes from the Latin word, *digitus*, which may be traced back to the first century BCE, when it meant "finger" or "toe," or simply the act of counting (as in "counting on your fingers," because numbers under ten were counted on the fingers). In a letter to his friend Atticus, written between 68 and 44 BCE, the famous lawyer and scholar Cicero complains about how Brutus (Atticus's friend) wants to charge him 48 percent interest, instead of the usual 12 percent. Offended, Cicero says, "What a wide difference this implies you will certainly be able to reckon, if I know your fingers." This phrase, "if I know your fingers" (*si tuos digitos novi*),[41] implies that the use of one's fingers to count when carrying out more advanced operations, such as calculating of interest and debt, dates to at least 68 to 44 BCE. The fingers were used as concrete, physical units of measurement—to count on and point with—and also facilitated symbolic operations that enabled the negotiation of prices, deals, loans, and so on. The phrase, "if I know your fingers" may be understood to reference a more general, common method known and used by all: that is, a sort of universal system where operators and functions were rather fixed.

The term *digital* may be traced to the proto-Indo-European root form, *deik*, which signifies "to show." This root, highlighted in the *American Heritage Dictionary of Indo-European Roots*, is connected to various words, including the Sanskrit *dic* (meaning "to show" or "to point out"), Greek terms such as *deiknynai* ("to show" or "to prove") and *dike* ("custom" or "usage"), Latin words such as *dicere* ("to speak, tell, say") and *digitus* ("finger"), and Old High German *zeigon* and German *zeigen* ("to show").[42] This connection, related to a word extensively used in societal, economic, juridical, cultural, and aesthetic contexts, is significant. Beyond its common usage, the *deik* root conveys an alternate interpretation of digital—one that transcends mere registration or storage of binary-encoded information.[43] Instead, it reveals a perspective that emphasizes the contextual representation of the world: it indicates the contextual presentation of the world, that is, to the concrete subjects and objects whose stories are told, and to the aesthetic and cultural situation in which they take part. Conversely, the *deuk* root, also of proto-Indo-European origin, has a different focus. It indicates the potential for "pulling," "dragging," or "leading something away." This root underpins essential scientific concepts centered on the notion of "leading away" or "deriving." Concepts such as reduction, induction, deduction, and abduction, integral to computer science and information-oriented research, gain prominence here. Notably, induction, deduction, and abduction have achieved distinction through their application in the work of philosopher and mathematician Charles Sanders Peirce.

In his Lowell lectures of 1866, "The Logic of Science; or Induction and Hypothesis," Peirce describes the complicated relations between objects and signs. He believes that logic, in the broadest sense of the word, is closely related to the ways in which things are represented, that is, with the general ways in which signs operate.[44] In his lectures, he connects two of his best-known triadic models: the three trichotomies of the sign—iconic (similar), indexical (causal), and symbol (conventional) representations—and his classification of reasoning in a schematic diagram that describes the fundamental relations between representamen and object. Abduction (which Peirce calls hypothesis at this point)[45] is linked to icons (through likeness), induction to indices, and deduction to symbols. He writes:

> We come to [. . .] the argument. [—] It will therefore be divided into three species according as this representation is a likeness, index, or symbol. These three species are the same as Hypothesis,

Induction, and Deduction. Hypothesis brings up to the mind an image of the true qualities of a thing—it therefore informs us as to comprehension but not as to Extension, that is it represents a representation which has Comprehension without Extension; in other words it represents a likeness.[46]

The representational relations between signs and things are fundamentally connected to the three modalities of the *deuk* root—according to Peirce, to abduction, deduction, and induction—and may be expressed as shown in table 1.

Abduction	Induction	Deduction
icon	index	symbol
likeness	nearness	convention

TABLE 1. The three modalities of the *deuk* root, as they relate to C. S. Peirce's three trichotomies of the sign and triadic classification of reasoning.

These modalities share the fact that their semiotic and representational qualities relate them to acts of derivation, etymologically marked by the *deuk* root. In other words, they illustrate an approach that derives by either explaining a new connection (abducing), generalizing from observations (inducing), or inferring from theory (deducing). When linked to the three trichotomies of the sign, these derivations may be expressed a little differently, yet still focus on their inferring role in reasoning: here, the interest is directed by relations of likeness, for instance, between a thing and an image (icons), relations of causality between something already known and its traces (indices), or relations of convention between something observed many times and its possible meanings (symbols). All these modalities derive something from the things they describe, and may again be connected to the deuk root form, and the technical or informational interest in derivation, that is, in the relations between representation and thing in counting.

An approach based on the alternative etymological connotation of the *deik* root form disrupts the meaning of digital. Here, the focus is less on "leading away," or "deriving," and more on "showing," for instance, what happens, the ways in which it is experienced, and the position from which something is seen. This difference may be tentatively visualized as shown in table 2.

The *deuk* forms			The *deik* forms	
Abduction	Induction	Deduction	Deixis	Pointing to or specifying the function of words, whose denotation changes from one discourse to another
icon	index	symbol		
likeness	nearness	convention		
				Center of experience
			Digital	Counting on your fingers
				"this" or "that"
	Sign level			*Process level*

TABLE 2. Schematic visualization of *deuk* and *deik* forms as they relate to C. S. Peirce's three trichotomies of the sign and triadic classification of reasoning and the etymology of the word *digital.*

Although the *deuk* forms are built on derivation, that is, an approach that extracts and "leads away"—operating on representations of the subjects studied—the *deik* forms are built on presentations, that is, on an approach that focuses on the processes and contexts that include the subjects. Translated into a digital vocabulary, a shift from representation to presentation marks a shift in interest from a focus on formalizing or abstracting the meaning of digital phenomena, that is, from their symbolic value, to a focus on displaying, presenting, or contextualizing how they operate more generally, for example, culturally or aesthetically.

Although a focus on the abstract, informational level of representation is both relevant to and necessary for many professional disciplines, a more process-oriented concept may help to provide new analytical insights. Instead of focusing exclusively on digital phenomena as logical-mathematical and informational, derived from the surrounding world as sets of data, they may also be approached as something to be perceived with the senses, as something to be experienced in a certain way in a specific setting. When Cicero counts on his fingers, finds that the interest is too high, and complains to his friend Atticus, it is precisely this choice of approach that is evident: the symbolic universal system, which they both use, may presumably be analyzed on the basis of its logical-mathematical qualities as a method for calculating interest, yet it may also be discussed in relation to the situation in which Cicero finds himself, as he looks at his fingers and does not understand why they—these fingers that usually provide an accurate method of counting, that usually provide him with a way to calculate, discuss, and visualize all kinds of mathematical matters—suddenly fail him. The universal system is a method that, like so many other things, is restricted to the context in which it operates, a context that makes some calculations possible while excluding others, that deceives one part while letting the other win.

DEICTIC SITUATEDNESS

One of the perhaps best-known applications of the *deik* form is found in the concept of *deixis*. In linguistics, deixis is understood as an encoding of spatiotemporal context and subjective experience of an utterance. Words such *here*, *now*, or *this* are examples of so-called pure deictic terms, as they depend on context and the speaker's cognitive center of orientation.[47] For instance, in the sentence, "let's meet here tomorrow," an understanding of what *here* refers to is decisive for its making sense. *Here* becomes an indicative word, and the etymological root, *deik*, is apparent in the denotation of the word: deictic terms point to or *show*. Deictic words such as *here* or *this* need a context and a center of orientation to make sense. Deixis needs context, you might add.[48] Although the *deuk* forms (induction, deduction, reduction, etc.) abstract and formalize, and in this way lead away from the context and situation of the linguistic exchange, in contrast, the *deik* forms depend on a context's situatedness. The etymological connotation of *digitus* described above indicates this less abstract understanding of the term, that is, the spatiotemporal context and subjective experience. Although this connotation has been overshadowed by technical and informational understandings of the word *digital*, it is still possible to trace back to this etymological alternative, and also, in more concrete ways, to use it to think with.[49]

This book aspires to follow a more deictic approach by describing the actual experiences and contexts that form the basis of cultural techniques of counting and sorting. It dives into the particularities of digital culture: into the concrete practices of data processing as they manifest in census-taking, punched-card operations, statistical summaries and models, face recognition aggregates, and more. Just as Peirce speaks of a "thirdness" or middle of communication, this book aspires to begin from the "middle of things,"[50] rather than "from above."[51] Therefore, this book's descriptions of the central term *digital* are made with concrete digital operations and their relational (social, cultural, and aesthetic) influence in mind. Thus, *digital* has two meanings: it identifies the specific contexts and situations in which digital calculations are actualized and—at the same time—synthesizes the disparate array of ways in which this actualization is carried out. It covers a range of ways in which discontinuous data are displayed and described by statistical concepts and visualizations, and the particular, theoretical mindset or way of thinking that underpins these expressions. Whether they are mathematical, statistical, or philosophical, theoretical maneuvers always presuppose an object of thought. The central object of thought in this book—the digital object—is not the pixelated representation on the screen or the lithium battery that supports its appearance; instead, it is a new type of object that finds its way into everyday life, into political decision-making, into research politics, into social spaces, and into aesthetics. It is not a new class of objects that needs to be covered or schematized, but a reaction to a need to reflect on the digital operations that renegotiate so many things around us, from the largest, general structures of global politics to the smallest, most intimate relationships.

THINKING WITH CONTEMPORARY ART PRACTICE

Though it might not be the obvious choice, given the theme of statistical and data-driven knowledge production, this book incorporates concrete artistic projects and experiments as fundamental components of its analyses. However, it is important to underscore that these projects are not intended to be the primary empirical material of the book. Rather, they are used as tools for thinking about the various themes addressed throughout the chapters. Incorporating Mieke Bal's understanding of "travelling concepts" and drawing inspiration from her efforts to let her objects of analysis "speak back" to the analytical concepts,[52] the intention is to make room for precisely the *artistic* form of reflexivity. Conceiving artistic practices as dialogical partners rather than fixed objects of study, the book aligns with and is a contribution to the field sometimes referred to as artistic research, an approach which distances itself from ideas of "artworks" as something objectively defined, in order to instead constitute a form of "reflective transformation" of an otherwise nonartistic lifeworld, as Juliane Rebentisch calls it,[53] offering a particularly artistic form of knowledge production to supplement more traditional, scientific "ways of knowing."[54] Thinking *with* or *along* art is thus an approach that runs throughout the book's chapters, and which is reflected in its concrete interpretive practices.

THE STRUCTURE OF THIS BOOK

This book is composed of three main chapters, each consisting of a conceptual study of a particular digital mode of operation, represented by the statistical entities of outliers, aggregates, and patterns. Each chapter is based on close readings of a selection of contemporary art projects.

The first chapter focuses on the statistical outlier and the data set values that are intentionally excluded and marginalized. The outlier, as figure, is addressed on two levels: first, on a concrete level, where the historical application and understanding of data processing is studied, and secondly, on a more general level, where exclusion is understood culturally, politically, and aesthetically, based on omissions of context and metadata, or of certain historical events or individuals, for example. In this chapter, a central argument is that outliers play a decisive role in the history of data processing and also in the history of the computer and the introduction of digital systems into modern society. Accordingly, this chapter explores, first, how concrete strategies for homogenizing data sets moderate and mark the study output and secondly, how exclusions, more generally and literally, make some forms of knowledge visible and others invisible. The central artistic exploration included in this chapter is the artist Rossella Biscotti's *Other* project (2014–2015), in which she focuses on the institutional dimensions of outliers (aesthetic, social, economic) and on the technological structures that facilitate their operations (machines, buildings, and techniques). Several large textile pieces produced on automatic Jacquard looms study and present the individuals excluded from comprehensive demographic surveys undertaken in Belgium in the first decade of the 2000s. By employing the textile's medium-specific character, Biscotti visualizes not only the embedded statistical outliers, but also the actual process that yields the censuses' specific results.

The second chapter focuses on the theoretical and historical idea of statistical aggregation as a mode of reasoning that selects, combines, and merges to make sense of data sets. In the nineteenth century, the word *aggregation* denoted the combination of observations, a definition that conveys the idea that there is a gain in information to be had, beyond what is revealed by the individual values of a data set, by combining them into a statistical summary. A central argument in this chapter is that combination, as a central mode of operation, contributes to understanding how the aesthetic processes inherent in presentations of data not only make it possible to manage the data sets, but also to make them visible in the first place. As such, the "data presentations" created by aggregation are described as a special kind of operative image.[55] Through an aesthetic close reading of Adam Broomberg and Oliver Chanarin's artistic project, *Spirit Is a Bone* (2014), this chapter focuses on the human face as it is translated into computational data. It thematizes the limitations of what a face is, as it challenges fundamental ideas of subjectivity: When does a face cease to look like itself? What are the most important characteristics of a face? What is noise, and how is one to identify the most essential parts of a face?[56]

The third chapter focuses on the philosophical, technical, and aesthetic idea of the pattern and with it, on more general discussions of so-called data-driven knowledge production[57] and techniques and methods, which convey the process of pattern recognition and abstract comparison as such.[58] This chapter argues that knowledge does not just emerge from analytically applied concepts or models,[59] or from the data set "itself,"[60] but emerges from a mediated interplay between context and conventions: concrete visualizations and diagrammatic abstractions.[61] With new methods for data processing that include machine learning and other types of "intelligent" systems, new statistical objects emerge, new classifications become possible, and new virtual connections are created. The third chapter's point of departure is Stéphanie Solinas's art project *Dominique Lambert* (2004–2016) and the numerous metatheoretical and methodological questions of pattern recognition, comparison, and comparability to which it gives rise. In this chapter, comparison is broadly understood in terms of discussions of the challenges and possibilities attached to ideas of comparatism as a methodological strategy in art and literature analyses and as a broader theoretical and scientific approach. The *Dominique Lambert* project questions not only how and when different things may be compared, but also how patterns and comparisons may be understood more generally, and how they are legitimatized in relation to humanistic knowledge production.

One

OUTLIERS

If they are not part of the durable furniture of the world, the same everywhere and always, statistical entities nonetheless change the world.[1]

••• Do averages exist? Are normal distributions real? What about the population of Belgium? This chapter delves into a pivotal concept used extensively, both in the field of statistics and in the sciences more broadly. Although many scientists take it for granted, this concept's epistemological status remains puzzling. This chapter zeroes in on the statistical outlier—the intentionally excluded and marginalized—for two key reasons.[2] First, the phenomenon of the outlier is intertwined with historical and current data processing methodologies. This includes how these techniques were and are applied to extensive data sets, such as statistical analyses of populations or socioeconomic studies. This specific context serves as the core focus for a detailed examination, revealing the intricate impact of a small set of analytical and material tools and inventions on today's understanding of data processing. The second reason for scrutinizing this mechanism lies in its connection to broader notions of exclusion—cultural, political, and aesthetic—for instance, the exclusion of minorities through omissions of context and metadata (e.g., by detecting outliers in demographic data sets) or the slow erasure of certain historical events and persons in written history. An instance of this would be the omission of the contributions of skilled workers in France and Great Britain's textile industries, whose crucial role in the Industrial Revolution is often eclipsed by enthusiastic descriptions of the technological inventions that pushed industry forward and paved the way for modern mass production.[3] Similarly, this chapter underscores the often overlooked female "computers" of the nineteenth century, who were responsible for complex calculations in fields such as insurance and investment, and also physics and astronomy.[4] The narrative of exclusion—outlying data points found in the tail of distributions—unveils a significant facet of the evolution of data processing and of the emergence of computers and digital systems in today's society. Throughout this chapter, I delve into the core mindset that is encapsulated by the concept of the outlier. I first examine the strategies employed to standardize data sets, which influence the outcomes of statistical analyses. Secondly, I explore how, in a broader sense, exclusions make certain ways of knowing visible and others invisible.[5]

This chapter's central thesis is that outliers transcend their mere mathematical or technical nature and prompt us to perceive them through the lens of their aesthetic manifestation and their role in mediating knowledge. These outliers are intricately intertwined with what I refer to as a digital or numerical mode of thinking—a spectrum of filtered forms of logic, categorization frameworks, and decision-making processes that help their creators to understand the data set to which they belong. It is crucial to acknowledge that this aesthetic character does not pertain only to communication or dissemination, as exemplified by data visualization or infographics.[6] It also encompasses the actual act of knowledge generation itself—the ways in which knowledge becomes visible and tangible.[7] This chapter focuses on the concrete ways in which modes of exclusion and omission occur.

ANATOMY OF AN OUTLIER

Concerns about inaccurate, contaminated, or deviating observations have long been acknowledged,[8] and the Gaussian distribution is not the only instance of a methodical approach to arranging and classifying observations based on a concept of normality.[9] Numerous mathematicians and researchers throughout the natural sciences have sought ways to interpret and classify outliers. Sometimes, this pursuit involves overtly rejecting outliers to maintain the integrity of an entire data set. Historically, disciplines such as astronomy, geodesy, chemistry, physics, and ballistics have grappled with the limitations of rejection and rules.[10] This interest in managing data heterogeneity also extends to domains such as national and socioeconomic contexts.[11] In extensive, systematic studies, noise and outliers pose significant threats, because they have the potential to introduce doubt and to discredit. Nonetheless, outliers may also offer opportunities for new discoveries and variations that automatic, unreflective rejections might overlook.[12] Thus, handling outliers straddles two divergent approaches, each with far-reaching implications. Opting to entirely discard an outlier in pursuit of stability and uniformity risks forfeiting crucial, valuable insights. Conversely, if all outliers are included, a data set may become contaminated with erroneous measurements, which obscure essential trends and correlations.

Today, statisticians and computer scientists embrace various types of outliers, because of their potential to offer intriguing insights, particularly in uncovering novel or unforeseen forms of variation.[13] Although outliers were once considered disruptive data elements that needed to be swiftly eliminated, modern outlier detection has evolved to be an important problem-solving subfield of computer science. Emerging domains of statistics and computer science, such as data mining and machine learning, are revisiting outlier-related challenges and introducing innovative methods for detecting and managing them.[14] This perspective will be revisited in the concluding section of this chapter.

The central focus of this chapter pertains to outliers in demographic studies. In this context, an individual's status is established through responses to a series of hierarchically structured, binary questions related to marital status. These questions include whether one is married, has children, or cohabits with a partner. Notably, the "other" category—the outlier—is the final alternative: a classification for those whose options have been exhausted. "Other" is assigned when an individual does not fit into the predetermined categories outlined by census administrators, or only aligns with a category of such limited size that it holds minimal practical significance in the broader census framework. As a label, "other" delineates a periphery that encircles the dense core of the data set, akin to the flatter tails of a Gaussian curve, or the least significant zones in the information space. Essentially, "other" represents nothing but itself. In semantic terms, it stands as an empty category employed by extensive data-mapping endeavors, owing to pragmatic considerations or expedience. It does not have any distinctive attributes, nor does it reference any actual frame of reference for the grouped data points. Unlike categories such as "married," "parent," or "single," the "other" classification lacks an inherent positive association and is primarily defined in negative terms—a pragmatic necessity to consolidate all outliers into a single category.

The foregoing perspective on outliers is particularly interesting when examined in the context of the artistic project *Other* by Italian artist Rossella Biscotti, which serves as the foundational inspiration for the concepts discussed in this chapter.[15] Executed between 2013 and 2016, Biscotti's project explored the anatomy of the outlier as figure. Through a methodical exploration that encompasses the outlier's institutional facets, from aesthetics and social dynamics to economic dimensions, on the one hand, and on the other hand, the technological frameworks that facilitate its operations, comprising tangible machinery, buildings, and techniques, Biscotti endeavored to chart the outlier as a conceptual entity.

In a series of expansive textile works meticulously crafted on automated Jacquard looms, she explored and presented individuals who were excluded from the standardized and preestablished categories introduced during Belgium's extensive socioeconomic mapping in the early 2000s. These individuals found themselves relegated to minority and accumulation classifications, owing to their failure to fit into the systematic logic of the classification algorithm. Each textile bore a subtitle that linked it to the informational conflict it depicted. For example, one of the textiles was titled "Dead Minorities," whereas another was designated "Single Mothers."[16]

By employing textiles' inherent qualities, both in terms of their medial nature and their ability to generate patterns, Biscotti achieved her twofold objective. She did not only make visible the diverse statistical outliers that had been disregarded as data points during the investigative process, but also illuminated the actual mechanism responsible for producing this outcome. To put this differently, her use of the textile as a medium went beyond merely presenting an array of data points that may have otherwise faded into obscurity or been ignored. Instead, she imaginatively incorporated weaving and the automated Jacquard looms' intrinsic binary pattern construction, effectively drawing parallels to the historical and current methodologies of data analysis.[17] Through her project, Biscotti adeptly showcased how the principles of algorithmic thought, as embodied in both textiles and the pattern-generating procedures of data analysis, possess a significant aesthetic dimension. This aesthetic facet assumes a central role in the project's overarching concept, notably embodied by the figure of the outlier.

Throughout this chapter, I am inspired and stimulated by Rossella Biscotti's artistic inquiries and the works she has crafted. These textile-based works catalyze my exploration of the evolution of today's digital data processing, with a specific emphasis on the modalities of exclusion. This pertains to instances where intricate concepts such as parenthood, gender, or marital status are distilled into data points that conform to the dictates of a data set. Subsequently, these concepts are separated, sorted, and treated in a manner that erases potential conflicts and dichotomies. In the preface to *"Raw Data" Is an Oxymoron*, Lisa Gitelman points out that this process involves systematic manipulation that may obscure intricate and important nuances.[18] The systematic modes of thinking and reasoning—what I refer to in the introduction as a distinct digital logic or operational framework—risk developing what Wendy Kyong Hui Chun terms a *habitual* quality.[19] These logics or modes of reasoning do not only operate and sort on our behalf through systematic and external maneuvers, but are also gradually internalized within our bodies, and act in ways that resemble those of what Michel Foucault and Giorgio Agamben identify with the term *apparatus* (or rather, *dispositif*)[20]—an invisible yet intricately interwoven set of mechanisms or structures that amplifies and perpetuates the power imposed on the social body.

INTERWOVEN EXCLUSIONS

At first sight, the construction of Biscotti's artworks may appear decep-
tively straightforward: a woven textile is affixed to a metal bar suspended
from the ceiling, one part hangs from a simple mounting mechanism,
and another part extends onto the floor. Beyond the figures and forms
that adorn the fabric's surface (the data patterns of the Brussels censuses), the textile is installed perpendicular to the floor, marginally
disrupted by slight wrinkles and creases in the soft material, as seen
in figure 1. However, these textiles transcend the role of mere picto-
rial planes; instead, they serve as components through which content is
communicated structurally.[21] The warp, conventionally an underlying
and often concealed framework of woven fabrics, has an active role as

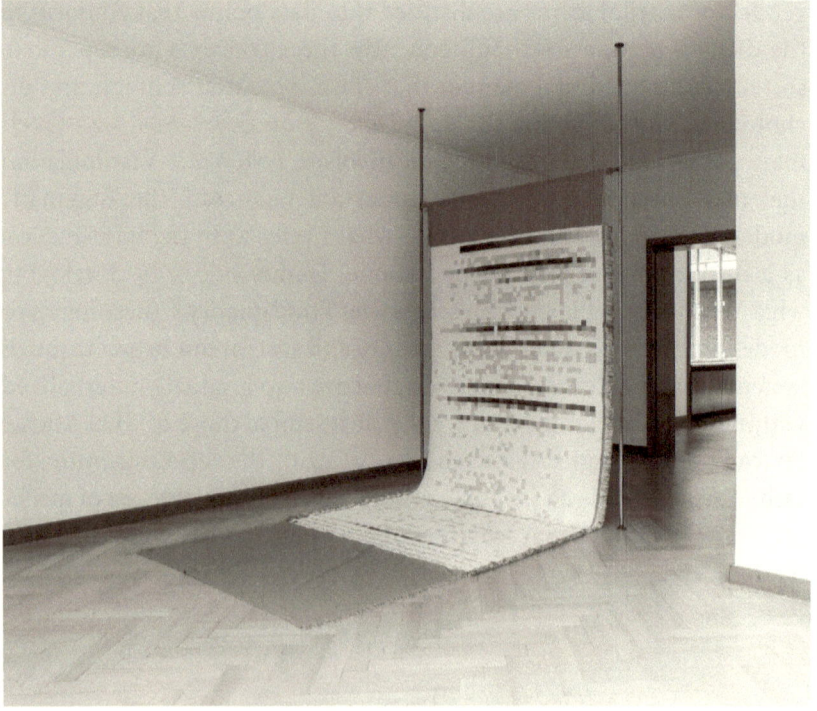

FIGURE 1. Installation view. Rossella Biscotti, *Single Mothers*, *10 × 10* at Kunstmuseen
Krefeld (2014) © Rossella Biscotti, Kunstmuseen Krefeld – ARTOTHEK.

a design element. This allows for not only the depiction of squares and hues on the surface, but also for the visual presentation of the medium's inherent binary structure.[22] Thus, the textiles are not static representations and assume an added dimension, where the structural intricacies themselves contribute to the narrative.

In conjunction with the grey squares woven into the fabric, the warp assumes a significant role in the artwork. The design aesthetically harnesses weaving's inherent potential. This mode of presentation unmistakably evokes the 1920s and 1930s, particularly the architectural landmarks of the Weimar Bauhaus movement, and various other sources of inspiration.[23] Several parallels spring to mind. One is the genre of abstract textile works from the 1920s, exemplified by the output of artists such as Swiss Sophie Taeuber-Arp, French Sonia Delaunay, and German Anni Albers. Each adeptly manipulated the textile's structure, capitalizing on the binary interplay of warp and weft to propel what Markus Brüderlin terms the "birth of abstraction from the spirit of the textile."[24] Similarly, there is a correspondence between the configuration of warp and weft and the horizontal-vertical composition inherent in the image's surface. This interplay has been likened to the rectangular grids found in paintings from the early 1900s,[25] as highlighted by Rosalind Krauss in her seminal 1979 essay.[26] The works of artists such as Piet Mondrian and Kazimir Malevich, characterized by their use of grid-like structures, reflect this composition. Consequently, these textile surfaces have a distinct haptic quality and evoke a sense of visual "touch" when viewed, effectively rejecting conventional perspective and depth.[27] Furthermore, the setting of Biscotti's artworks resonates with modernist echoes. Their exhibition at the Mies van der Rohe–designed villas in Krefeld emphasizes this connection. The villas, now a museum space, showcase a design ethos akin to that of the square modules featured in the textiles. Just as these modules configure the notation in the textiles, Mies van der Rohe employed an interlocking cube system to design the volumes of these houses.[28]

Several factors come together to imply a modernist backdrop for Biscotti's artworks. It is alluring to analyze the series of textiles through this lens, appreciating their abstract expression and references to a period marked by the redefinition of both textile production and artistic explorations through mechanization and rationalization. Nevertheless, an alternative perspective will thread its way through this chapter: Although Rossella Biscotti's *Other* series undeniably incorporates unmistakable references to the modernist aesthetics of the early twentieth century, I suggest that these works should be interpreted through a *digital* lens. First, their digital nature may be attributed to their intrinsic digital *poēisis*, or their actual method of creation.[29] Second, this assertion may be expanded to contend that the works also address central inquiries into the essence of digital phenomena on an aesthetic level. By engaging the senses (*aisthêsis*),[30] they prompt a contemplation of our perception and comprehension of the nature of digitality.

A closer examination of Biscotti's artworks reveals a discernible systematic arrangement of patterns—comprising grey and colored squares—on the fabric's surface. The pieces draw not only on abstract or modernist aesthetics, and capitalize on the distinctive structural characteristics of the medium and materials employed, but they also have a representational dimension that hints at something beyond the artworks themselves. For instance, a deeper scrutiny of the piece *Other (Acquired Nationality)*, seen in figure 2, emphasizes the importance of the woven legend at the textile's edge. This legend assumes a pivotal role in unraveling the piece's meaning, as it offers tangible information about its content.

Each shade of grey in the textiles has a specific denotation: they signify the frequency with which *something* occurs, or how a distinct quantity of *something* is distributed over a range of data points. The textiles' assortment of differently colored squares conveys varied values, decipherable with the help of the key presented in the legend. The textiles transcend mere visual planes and evolve into intricate graphs that encapsulate data. Meticulously color-coded, they stand ready to be decoded, effectively mirroring the language of data representation.

The exhibition catalog includes preliminary studies for Biscotti's textiles.[31] She undertook various forms of data processing prior to beginning the computer-operated Jacquard weaving. In the case of *Other (Acquired Nationality)*, values are summed, based on two distinct categories, and presented in a table in the catalog. These numerical values were translated into frequency, represented by the subtle variations of

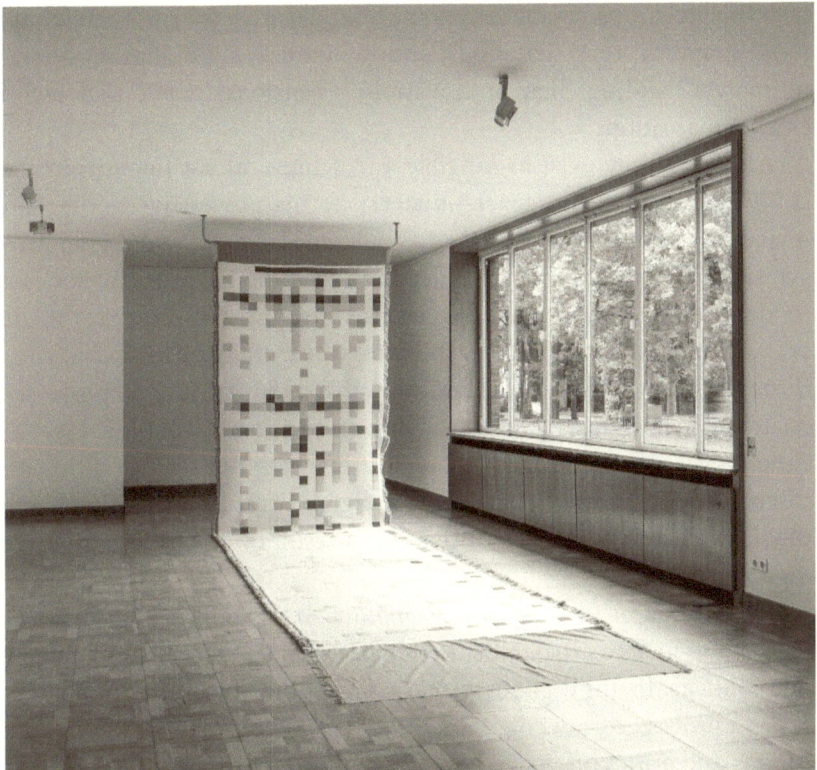

FIGURE 2. Installation view. Rossella Biscotti, *Other (Acquired Nationality)*, 10 × 10 at Kunstmuseen Krefeld (2014) © Rossella Biscotti, Kunstmuseen Krefeld – ARTOTHEK.

grey in the squares. A yellow and a red square are designated as statistical markers and assigned attributes such as name, gender, age, marital status, nationality, and the residential district of an individual. Additionally, they are linked to the so-called Household Identification Numbers assigned to anonymous citizens.

The patterns of the textiles serve as encoded depictions of a preliminary diagram that predates the weaving process. This diagram comprises x and y axes, graphically visualizing anonymized demographic data extracted from the 2001 Belgian census, linked to Belgium's National Register.[32] Consequently, the textiles' tactile surfaces relate closely to their graphical blueprints, which intertwines them with the digital procedures to which the material manifestations allude. This

entwined relationship challenges the boundaries between tangible surfaces, graphical representations, and the underlying digital processes,
ultimately melding them into a multidimensional whole that defies
clear demarcation.

Although they exist as tangible and sculptural entities within the
confines of a museum space—material forms possessing tactile, soft
imagery—these artworks emerge as outcomes of a defined digital progression. It is important to understand this digital process, not just
to identify and categorize the series as an example of computer-made
art,[33] but also to conceptualize the central reasoning or logic that runs
through the artworks, and accordingly, through many digital processes
related to census-taking, in particular, and national economic investigations of demographics, in general. It is worth noting that although
Biscotti's series is not presented as a digital display in the immediate
physical sense, it is important that it is a result or a manifestation of a
digital process.[34]

In his book, *A Philosophy of Computer Art*, Dominic McIver Lopes
brings into focus a crucial distinction between two facets of digital art:
one being art that is digitally *displayed*, the other encompassing digital
art as a process, termed digital *technē*.[35] Lopes notes that many art historians and critics often overlook the significance of the digital process
that underpins the creation and activation of digital artworks. These
processes, although not inherently destined for digital display (e.g., on a
screen or as part of an interactive installation), remain pivotal in shaping our understanding of these artworks:

> Digital art is either made by computer or made for display by
> computer in a common, digital code. The digital display is obvi-
> ous enough when we come across it, and so is its impact, which
> includes new opportunities for multi-media and new venues for
> audiences to access art. A less obvious "game changer" is the use
> of digital encoding to make art.[36]

If one follows this argument, one might ask: What is the nature of the
digital process used in Rossella Biscotti's artworks, and could this process hold the key to understanding them? As previously mentioned, the
data points represented in the fabric reference the national censuses

conducted in Brussels in 2001. This highlights not only the contemporary relevance of the digital process that underpins these artworks, but also the historical interplay between now-obsolete technologies, such as punched cards and the enduring binary logic that forms the foundation of the systems governing demographic politics. Therefore, a pivotal component for understanding these artworks is the set of preliminary investigations Biscotti undertook prior to creating them. These inquiries examine the diverse forms of data processing intrinsic to census-taking, both past and present. In these studies, Biscotti unveils the intricate evolution of data processing itself, tracing its development from early tabulating machines and punched-card systems to more recent counting systems, while exposing the persistent errors that have been carried along across generations. In the exhibition catalog, Adam Kleinman, the curator, aptly notes that this parallel resonates with current developments, particularly considering the current dominance of Big Data systems. These systems incorporate not only colossal volumes of data, reaching nearly incomprehensible scales, but also inherit the logic of sorting and organizing data patterns, a lineage that extends to the foundational systems developed in the eighteenth and nineteenth centuries.[37]

CENSUS-TAKING ON THE MARGINS

Discussions about the census-taking process and the fundamental logic of data processing serve as a fundamental aspect of Biscotti's artistic pursuit. In the exhibition catalog, and particularly in a short paratext coauthored by Biscotti and statistics and computer-science experts, an exploration unfolds the technical intricacies that underpin the production of these works.[38] In a census, an individual's societal status hinges on their responses to a hierarchical flowchart featuring "yes" or "no" queries that revolve around the criteria of the conventional family unit. Figure 3 presents this scenario, where questions such as "Do you live alone?," "Are you married?," and "Do you live with a partner?" are common inquiries that contribute to determining an individual's standing as a counted member of society. The designation "other" represents the last option—a classification reserved for individuals who

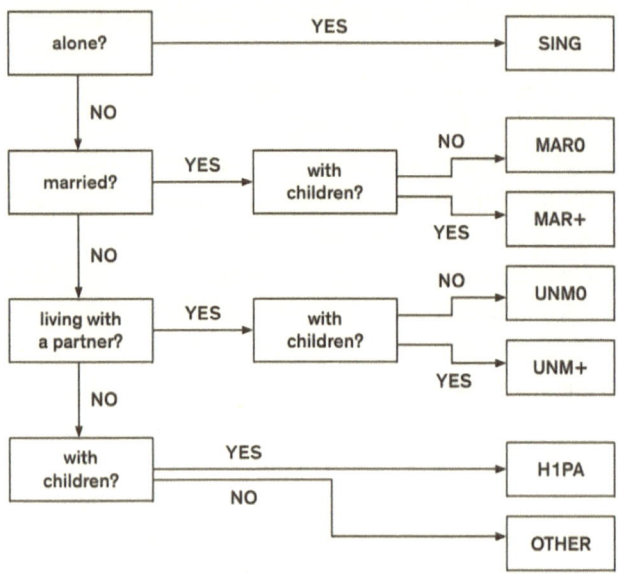

FIGURE 3. Flow chart used in the 2001 Brussels census, and later printed in the exhibition catalog for the *Other* project at Art Museum Krefeld. The category of "other" in the lowermost corner is the last into which one may fall, for those people who do not fit into any of the other predefined categories. © Rossella Biscotti.

have, in a sense, fallen outside the spectrum of all the positively defined possibilities provided, captured by labels such as *MARo* or *MAR+*.[39]

Individuals who share their living spaces with fellow students or colleagues, who reside in communal or other cohabitation arrangements, or who are temporary tenants or grandparents who have opted to live with younger family members find themselves in circumstances that defy neat categorization of conventional household designations such as "single" (*SING*), "married, without children" (*MARo*), or "married, with children" (*MAR+*). The ordinary categories inadequately capture the diverse living arrangements and relationships prevalent in contemporary society. However, Rossella Biscotti's project does not limit itself to patterns prevalent in conventional family and housing structures. Instead, it delves into the interstices inhabited by demographic outliers—those individuals who cannot be definitively encapsulated by the binary responses of "yes" or "no," "1" or "0" to formulaic questions.

Her endeavor centers on unraveling and comprehending the factors that lead certain individuals to become outliers in demographic studies and probing the algorithmic mechanisms that underpin the emergence of these distinctive categories.

In a tangible sense, the textiles in the project collectively shape a matrix composed of square modules. In this matrix, the x axis represents distinct, predefined conditions, whereas the y axis captures the number of individuals who fit those same criteria. These numbers are presented in grey tones created by different woven elements, where black depicts the highest density of people and white depicts the lowest, as seen in figure 4. Subsequently, some individuals are highlighted in the fabric, their data translated into text at the bottom of each matrix (and later, the textile), where an inscription clarifies that person's status.

Suspended from the ceilings of Mies van der Rohe's tight, functionalist cubes, the textiles seem to ask for interpretation in the historical context in which they are intricately woven: a history that revolves around those designated as "the others"—individuals omitted from

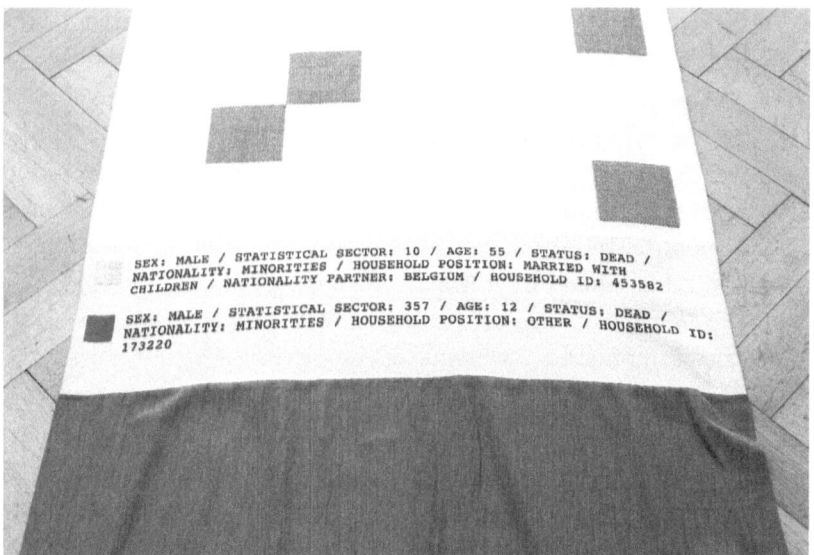

FIGURE 4. Rossella Biscotti, inscription detail from the piece *Dead Minorities*, *10 × 10* at Kunstmuseen Krefeld (2014) © Rossella Biscotti, Kunstmuseen Krefeld – ARTOTHEK.

comprehensive demographic studies, those who diverged from conventional familial structures. These textiles weave narratives of an era when modern architecture championed airy, minimalist spaces—a period when open layouts with sparse walls symbolized the dawn of a new era. However, this vision coexisted with more traditional developers' persistent demand for established, recognizable spatial divisions, spaces into which they could retreat and shut the door on the family sphere.[40] The textiles tell of a prosperous textile industry, since the widespread implementation of automatic Jacquard looms in European weaving mills enabled industry leaders to erect imposing villas designed by the era's most famous architects. Yet paradoxically, the same industry left its weavers empty-handed, as their manual craftsmanship progressively yielded to automation by "digital" looms.[41]

Biscotti's textiles have a dual character and embody the synthesis of material and metaphorical qualities. Their appearance encapsulates both the pliability of the fabric—evident in the wrinkles and undulations—and the structured rigidity of the square composition that presents a system that is far from malleable. This duality mirrors the architectural setting of Mies van der Rohe's villas, where the visible steel framework of the structural elements reveals a narrative of the building's own creation. Similarly, Biscotti's textiles bear witness to a binary framework, their deliberate accentuation evident in the double-weaving visible at the fabric's edge. Through this technique, the textiles do not only capture and highlight their visual nature, but also accentuate their distinct mode of operation. As a visual medium, the woven textile supports the fundamental logic that encodes its "data input," maintaining its binary organization through a continuous interlacing of under-over, 0-1, turned-off-turned-on. This process echoes the functioning of computational systems, effectively bridging the artistic and technological realms.

As a medium, woven fabric adeptly echoes the digital and systematic processes that yield it by employing one of its most distinctive attributes: the tangible interlacing of data, achieved through the fabric's intertwined threads. This medium bears the weight of a historical legacy characterized by deliberate, unhurried craftsmanship. Traditionally, it reflected a meticulous, if somewhat slow, process, which interwove discontinuous threads, materials, and, subsequently, patterns, through hours of apparently monotonous, physical labor.

The textile's image plane and its encoding of a binary structure—which conceptually intertwine in the *Other* project—are both supple and firm. On the one hand, Biscotti's textiles emphatically emphasize the dismissed and marginalized data points, giving them a newfound "voice" and an alternative mode of representation; on the other hand, they also exude an aura of latent frustration, prompting questions that challenge established norms: Do conventional family structures indeed persist with such *habitual* regularity that we remain oblivious to their role in shaping and cocreating the classifications employed in demographic studies? Is it feasible to continue to use the system's single "other" category to group together individuals who share nothing beyond their outlier status? Do systems such as this one inevitably *generate* these "outsiders"—individuals whose data are excluded because of design constraints, or dismissed as mere noise, or even misinformation—rather than simply *documenting* them?

Examined through the lens of the digital process, the textile reveals itself as far more intricate than a mere arrangement of intertwined or fragmentary data points. It emerges as a vehicle for a specific way of thinking—a particular *form* through which the unseen becomes visible. In the context of the *Other* project, weaving transforms into a technique for creating patterns and forging connections that otherwise would have remained concealed—a conduit for manifesting a general line of thought, you could say. The act of weaving assumes the role of a cultural technique, which forms and changes the materials it incorporates. The textile serves as a medium of translation, capable of transmuting the fragmented and disconnected into a unified and cohesive whole—a process of transformation akin to rendering the incoherent comprehensible and the disparate homogeneous.

WEAVING AS A CULTURAL TECHNIQUE

Like other techniques of representation, such as archiving, counting, and measuring, the process of weaving may be examined from both historical and conceptual standpoints.[42] This entails exploring how, in the *Other* project, the technique of weaving functions as a way of regulating and organizing what it conveys. Who is given a voice, and who is left mute? What kind of *thinking* does the process of weaving prompt? Considered as a cultural technique, weaving essentially generates patterns. It intertwines threads, strings, or ribbons to establish connections where none previously existed. It takes binary input, a range of components, and assembles them into a coherent whole, which may then be disassembled and readied for new configurations. Weaving, in the *Other* project, involves a deliberate selection process: its components are rarely arbitrary, and their sequential arrangement is important. Weaving combines the right elements, at the right places—it strategically interconnects its components. These need to not only align with the intended pattern, but they must also be *connectable* in the first place. As a mode of operation, weaving excludes as much as it includes. Every thread, every data point, must be meticulously prepared for the loom: the threads must be of the right thickness, colors must relate to each other, and the pattern should be adapted to processing by a binary system.

The traditional practice of weaving, predating the introduction of automated looms in the early eighteenth century, exhibits notable similarities to today's use of digital looms and their underlying programming logic. The fundamental technique remains the same: first, the weaving material is prepared; second, the pattern is designed by distributing the warp correctly; last, the process is executed systematically to achieve the intended result. However, from the perspective of cultural techniques, weaving also has a history far more intricate than the conceptual aspect of this process. Among the early instances of algorithmic processing,[43] there was an intriguing development in the 1720s, with the introduction of a paper roll to complement the drawloom.[44] This roll of paper was manually perforated in sections, each corresponding to a punch on the loom or a thread. Basile Bouchon, a textile worker from Lyon, initially devised this system, which was later refined to some extent by his assistant, Jean-Baptiste Falcon, fully automated by inventor Jacques de Vaucanson in 1745, and ultimately assembled and promoted by Joseph Marie Jacquard.[45] It is worth noting that Jacquard's machine, an iconic creation, is sometimes even regarded as a precursor to the "first" digital loom.[46] Jacquard's innovative loom entered production in 1801, and soon after, he patented it under his name. These automated, binary-encoded looms surpassed the productivity of the drawloom, a well-known device.[47] When coupled with the steam engine, Jacquard looms became an integral part of the swift technological advancements in manufacturing, which would subsequently redefine the landscape of modern mass production.[48] Whether an investigation examines weaving techniques or census procedures, the core principle remains the same: the interplay of under-over, 0-1, turned-off-turned-on. The punched card, as an object, proves equally versatile for capturing demographic trends and textile designs.

The possibility of altering a textile's pattern by merely changing the punched cards that encode thread positions is often recognized as a significant conceptual precursor to the evolution of computer programming. However, this progression toward today's automated systems involved numerous iterations and refinements, some of them related to the darkest chapters of the nineteenth and twentieth centuries' history.[49] Both conceptual and historical parameters are at work in what may be termed the cultural technique of weaving. Thus, weaving, like various other cultural techniques, occupies a space between process and object. It straddles habitual and regulated patterns that have evolved and consolidated over time, and more thing-like, well-established forms that, via technological maneuvers, for example, become aesthetically and materially substantial.

The focus on the technical and structural aspects of cultural phenomena, as found in the theories of cultural techniques, especially in the works of Sybille Krämer and Horst Bredekamp, aligns with the overarching goal of this book. This goal is to comprehend fundamental mathematical and philosophical ideas and concepts that have been devised for, and are integral to, modern data processing. Despite the predominantly theoretical nature of these concepts—whether they are products of invention or discovery, real or metaphysical—nonetheless, they change the world, as Lorraine Daston notes.[50] Krämer and Bredekamp emphasize that the possibility of changing a scientific discipline is closely related to the aesthetics of concepts, it employs. They assert that "[t]he eye of the mind is anything but blind," and elaborate, "it is precisely the sensualization—the aestheticization—of invisible processes and theoretical objects that are the fuel of scientific change."[51] Representation alone—the binary operationality of the loom, which encodes, defines, and organizes the patterns and figures in the textile—should not be isolated from the visual outcome or aesthetic interface[52] that is simultaneously produced. In Krämer and Bredekamp's words, "[i]t is in the (inter)play with language, images, writing, and machines—in the reciprocity between the symbolic and the technical, between discourse and the iconic—that cultures emerge and reproduce."[53] Thus, the modal operations of the loom, understood as a kind of cultural technique, are characterized by a symbolic and technical, that is, discursive and iconic, *reciprocity*, a reciprocity that reveals its aesthetic meaning. Krämer and Bredekamp state, "Visuality is anything but a merely illustrative sideshow—it constitutes the irreducible center for the research and evidentiary context of the sciences."[54]

In parallel to the way in which cultural techniques are understood and described as *medial* forms of agency,[55] I also attribute a pivotal visual and aesthetic character to the digital processes that I am investigating. Although Krämer and Bredekamp delve into the realm of computer-related cultural techniques, they do not speak of a decidedly digital operationality. Drawing on the example of infinitesimal calculus, they highlight the impact of computational and algorithmic systems. They emphasize how the manipulation of calculable symbols has given birth to new theoretical objects, including the evolution of the number zero as a case in point, but also mathematical objects such as differential equations or imaginary numbers. They write:

> On the one hand, the aesthetic of calculus is such that it "feeds" entities into the register of sensory perception that would otherwise be cognitively invisible; at the same time, however, such an aesthetic produces and constitutes these kinds of "objects" at the moment of their visualization in the first place.[56]

Introducing mathematical and theoretical concepts provides the capacity to visualize elements that might otherwise remain invisible and incomprehensible. However, this process also contributes to the broader impact on knowledge production through the process of externalization. An illustrative example of this relationship is seen in the functioning of digital principles in automatic looms. In this context, the binary system—representing under-over, 0-1, turned-off-turned-on—is not only enabled by abstract operations embedded in a physical substance, but it also becomes visible and real through the creation of patterns: the woven fabric reflects the digital, systemic process through one of its fundamental characteristics—the interlacing of data and physical threads in the textile. Similarly, Friedrich Kittler highlights the intricate connection between the communicative aspects and the productive capabilities of media. He suggests that media do not only shape the way in which communication occurs (representation), but are also the product of that very communication (storage medium).[57] Sybille Krämer supports this notion by noting the formative influence of mathematics on culture, which she sees as indicative of a symbolic-machinic relationship—essentially, two sides of the same coin.[58]

The idea of cultural techniques enriches this discourse by pinpointing and emphasizing the combination of aesthetic and technical-mathematical elements in the binary weaving process. It defines the medial agency, so to speak, of what Cornelia Vismann calls the medial *grammar*.[59] This overarching attention to the technologically mediated verbs of culture aligns well with my approach, provided it complements content analysis with a focus on the actual creation of meaning through activities, or "ways of *doing*," as Liam Cole Young terms it, that is, counting, measuring, collecting, observing, playing, confessing, and listing.[60] To grasp and develop an understanding of Rossella Biscotti's artworks and, by extension, to delve into pattern recognition in both manual and digital weaving, along with its broader connection to binary code, I propose integrating Krämer and Bredekamp's notion of cultural techniques with the narrower concept of the theoretical object. Although examinations of cultural techniques include both material and historical considerations, the epistemological figure itself—the object—serves as a tool for thinking. This tool comes into play when conducting theoretical work, sharing scientific insights across various knowledge communities, and even in contexts such as political and economic interactions.

Biscotti's works go beyond embodying just one type of operationality. They also realize a more general object that takes pattern generation as its object: in this context, patterns may be understood as indicators of a more broadly encompassing digital logic or functionality, a manifestation of something more overarching, which facilitates and prompts differentiation and exclusion. By calling it an object, I do not refer to the physical configuration of the pattern, encoded in it, such as a leaf, or the rigid, perforated cardboard of a punched card. Instead, I refer to the theoretical foundation that guides our thinking. The patterns serve as essential, albeit ephemeral, points of reference for digital operationality—an embodiment of ongoing knowledge production. This object operates and intervenes on a theoretical level but, nonetheless, is realized in concrete, aesthetic ways. As an object, it has partial visibility as a concrete pattern on the textile's image plane—a leaf or simply a connecting line—yet it cannot be reduced to this manifestation. The pattern formation is the result of a digital way of thinking that combines and organizes, and even though the outliers created by this process are not physically there, nonetheless they intervene in their own virtual, theoretical way. The concluding segment of this chapter will delve into the concept of the outlier as a quintessentially digital concept.

EMPTY STAND-INS FOR EXCLUSION

> *"Other" is the last box: a category of people that have "fallen out" of all given possibilities ... Investigating this label, we focused on different groups that contained one or more relationships or positions defined as "other," tracking different relational behaviours that run through the census as a pattern.*[61]

Rossella Biscotti characterizes the outlier as "the last box," intended for individuals for whom no other category applies. As a general figure that twines around the demographic study's many questions, the outlier has a pattern, just as do "single," "married," and "parent." However, the difference is that the outlier is nothing but a negative affiliation, comparable to other computational designations—such as *NaN* (not a number), a label for undefined or unrepresentable numeric data types, *NULL*, a pointer that indicates missing data in a database, or *default*, a computer scientist's term for an error or an as-yet-undetermined value—"other" denotes something that either resists categorization or arises unintentionally. As part of the bureaucratic classification system, the "other" category is nonetheless crucial, because it allows everything covered by a systematic study to be represented and thereby included.

In their seminal work, *Sorting Things Out: Classification and Its Consequences* (2000), Geoffrey Bowker and Susan Leigh Star define traditional classifications systems as follows: "A classification is a spatial, temporal, or spatio-temporal segmentation of the world. A 'classification system' is a set of boxes (metaphorical or literal) into which things can be put to then do some kind of work—bureaucratic or knowledge production."[62] In other words, classification must be complete to make sense,[63] and the set of boxes set up in the systematic map must include all the units of the study. Should an element not conform to existing categories, a provision must be made to allocate a designated space for it—often denoted as "other," "others," or "misc" (miscellaneous).

Bowker and Star state that the process of generating knowledge relies on a specific degree of virtuality. This requires using one or more stand-ins or technical placeholders that can accommodate less distracting or extraneous components of the study. For instance, in the context of a census, it is crucial to allocate a category that accommodates the final individuals counted, even if they do not neatly align with the rest of the system. Rossella Biscotti captures this situation as follows:

> The subject carrying the label "other" can be considered as default: a value representing nothing but itself, and positioning itself on the peripheries of the total or generalisation of a big dataset. The term "default" can signify two essential, contrasting meanings: 1) Default as failing or failure; omission of that which ought to be done. 2) Default as a value used when none has been given; a tentative value or standard that is presumed.[64]

Once again, the default takes center stage as a structured, numerical classification designed to accommodate values that defy conventional placement. In Biscotti's initial characterization of the default, individuals represented by groups of data points are perceived as *misleading* entities that do not fit into the predetermined framework and consequently highlight a sort of void, or domain of noninformation. In contrast, an alternate definition describes the default value as a preset or predefined value that is automatically assigned when no other viable values fit. In this context, the "other" class functions as a sort of aggregation space, which serves as a repository for the residual data points that may contribute to the study's information—data that are recognized but are not relevant to the broader context. The intricacy of this group is due to its characterization as a heterogeneous assembly that encompasses all those who do not conform, yet paradoxically come to share a commonality with one another—namely that of their categorization. Building on what Luciana Parisi, via Paul Virilio, has termed "negative optics,"[65] these individuals are aggregated based on their negative *differences*, rather than a positively defined *similarity*. These outlying "data points" are rendered noninformational by the classification scheme and are not only excluded from being counted in, as part of the study—forever residing in the tail of distributions—they are deliberately forced into a category of nonrepresentation, or of "non-value" and "non-mattering."[66]

Biscotti's delineation of the two default values may be correlated with two distinct philosophical perspectives on outliers, roughly associated with two distinct historical periods. The initial definition of Biscotti's framework echoes a prevalent idea from the latter half of the nineteenth century. At this time, the conceptual notion of the outlier was frequently associated with doubt and distrust, above all else. The American mathematician and astronomer Benjamin Peirce was a pioneer who developed an actual criterion for rejecting what he termed "doubtful observations," or outliers, in 1852,[67] and the American astronomer William Chauvenet revisited this argument in 1863. In systematic studies, divergent observations are deemed problematic because they may distort the overarching "image." The discourse surrounding Peirce's and Chauvenet's contributions sets the stage for more thorough scholarly investigations into outliers,[68] particularly concerning various modes of aggregation, a topic that will be further explored in chapter 2.

The second definition that Biscotti introduced resonates with more recent discussions of outliers in the realm of data science. Today, outliers continue to be understood as elements that come from the outside, indicating that the data diverge from, or significantly contradict, the rest of the data set.[69] Nevertheless, researchers are increasingly recognizing the potential of these aberrant measurements or data points. Currently, a more geometric awareness of outliers is gaining ground, as researchers approach data distributions by using spatially grounded metaphors.[70]

Outliers are no longer solely considered problematic or threatening due to their potential to distort or disrupt data. Instead, they are being formalized as a designated region of the spatial data landscape and are being understood in terms of dimensionality, rather than as mere noisy deviations from the rest of the data set. Currently, data scientists refer to methods for detection, rather than rejection criteria, and concepts such as novelty detection, anomaly detection, deviation detection, or exception mining[71] have a more mixed array of associations, compared to the discussions of noise and distortion from the 1850s and 1860s. Today, outliers are recognized as valuable components of the information space and are believed to contribute previously unrevealed insights into the realm of the rare, the exotic, and the unexplored.

The concept of the outlier has changed with the advent of Big Data, arguably influenced by the emergence of novel techniques from the field of machine learning, which help to reveal previously hidden patterns and correlations. Historically, the outlier denoted the unremarkable or, at best, marginally interesting data points, often on the fringes of relevance with respect to the primary investigation. Today, it catalyzes new research endeavors. Nonetheless, outliers continue to be defined in relation to the "norm," which pertains to the realm of the familiar, as opposed to the unknown, and the exceptional in contrast to the frequently occurring. Here, the virtual classification framework that Bowker and Star outlined resurfaces: aberrant measurements are an integral part of the information space (a fact that makes them particularly intriguing from a technical standpoint). They occupy a portion of the space with which we are already familiar, yet they diverge significantly from it in crucial aspects. Although they stand apart from the norm, they simultaneously contribute to defining this fundamental "normality" through their relative connection to the classification space.[72]

Returning to Rossella Biscotti's textile artworks, the dual nature, both material and metaphorical, of the marginalized "soft" data points in the data set and the artwork's "hard" expression of indignation may be situated in a specific context and historical framework. The outliers in Biscotti's works—visually apparent in the vibrant red and yellow figures amid a sea of subdued grey sections—may be aligned with Benjamin Peirce's concept of doubtful observations. These observations were dismissed because of their aberrant nature, as they failed to align with predefined categories, and thus found their place in "wastebasket" groups that encompass all that was deemed uninteresting.[73] Simultaneously, the artist's indignation is apparent in the persistent red and yellow sections, highlighting their resilience in the face of rejection. Rather than fixating on the recurring patterns of family structures, Biscotti's project specifically gravitates toward the interactions among the outliers that emerge from the demographic study. These interactions occur between individuals who could not give a simple "yes" or "no," 1 or 0, in response to the predefined queries. Hence, Biscotti's criticism is founded on a geometrical interpretation of the outlier—a conception rooted in a spatial arrangement of data points that gives rise to novel, connecting lines and that may reveal new insights into our individual and family lives.

A pivotal aspect of the *Other* project concerns the historical path of data processing, both in the past and the present, which becomes apparent through its technical aspect (reflected in the *poēisis* of the actual weaving techniques that reference the punched card of the Jacquard loom). This is further emphasized by their contextual role as historical and contemporary sorting machines, actively operating and intervening in political and social environments. As artworks, they perform the dual role of performance and narration, essentially chronicling the history of the outlier. This narrative unfolds through a blend of "soft" and "hard" approaches—by employing meticulous traditional weaving methods and contemporary systematic data processing strategies. Through representation and indignant presentation, they address the mode of exclusion. This mode, akin to the mode of aggregation discussed in chapter 2, encompasses an inherent aesthetics, which lists not only that which appears to be central or relevant within the logic of the system, but also that which is then determined to be peripheral or irrelevant with reference to predefined categories.

CULTURAL SORTING MACHINES

It is important to understand the ways in which knowledge is produced, both to use that same knowledge and to relate it to the complex context to which it belongs. Knowledge production is inherently situated[74] and concretely mediated[75]—as Kittler notes, communication is always entangled with production.[76] Rossella Biscotti's textiles reveal this through their material composition. They translate data into images—tentative amalgamations of patterns and figures. With her artworks, Biscotti explores not only the historical connection between the labor of weaving and demographic mapping, but also the fundamental process through which data are interwoven and "cooked" to make sense.[77]

Throughout this chapter, the woven textile has been my focal point, and it serves as both a medium and a mode, while I have also considered the intertwining of data points and their subsequent marginalization as a conceptual framework. By using systematic demographic studies as the principal lens, I have traced the role of the statistical outlier as it is present as a tangible entity in Rossella Biscotti's artistic work, and as an abstract statistical concept with historical and contemporary applications. I have demonstrated how the outlier functions as a category among many others in the field of demography. However, I have also highlighted its inherently negative and relative definition, which aggregates data points that deviate from the preestablished classification scheme and consistently reference the "normal" or "frequent." The term "other" represents nothing but itself and is a semantically empty category that has no distinguishing features and that provides no real frame of reference for the individuals to which it is applied.

The outlier, as a phenomenon, may be perceived as exerting a tangible influence on the historical application and comprehension of data processing. It continues to shape contemporary methodologies, especially the handling of vast data sets, such as those in demographic studies or socioeconomic analyses. However, its relevance is not confined solely to statistical applications; it extends to broader contexts of cultural, political, and aesthetic exclusions. On this conceptual plane, the outlier transcends its role as a specific phenomenon or a statistical concept. Instead, it emerges as an emblematic manifestation of a broader digital counting logic.

The final layer of my analysis corresponds to this book's core premise—an exploration of an inherently digital or algorithmic way of thinking. This overarching digital mode of thought, rooted in counting and exclusion, is embodied in the outlier. As a concept, exclusion operates on two levels: it refers to the specific contexts in which it appears and encapsulates a synthesized understanding of a disparate field of ways in which this realization shows itself. Exclusion is but one of many modes encountered in algorithmic thinking, just as the outlier is but one of many concrete ways in which this thinking reveals itself aesthetically and culturally. By digital counting logic, I allude in part to various ways of showing, portraying, and visualizing discontinuous data points as functions and patterns and in part to the special way of thinking that facilitates these kinds of theoretical operations.

Lorraine Daston writes that statistical entities such as outliers and averages may not be part of the durable furniture of the world, yet they wield an unsettling power: their artificiality makes them cohere or unravel, helps to develop hypotheses and explanations that may later be tested and further explored.[78] Yet this artificiality also helps them to intervene in, and form, what is visible and audible in the fields in which they operate: which help them to "distort, unfix, and inflect" the objects they describe.[79] In the next chapters, I further explore and unpack this question of artificiality, first through a study of aggregated facial images, and secondly through a discussion of comparison and comparability that comprises both traditional techniques of comparison and digital strategies employed in fields such as machine learning.

Two

AGGREGATES

He was, let us not forget, almost incapable of ideas of a general, Platonic sort. Not only was it difficult for him to comprehend that the generic symbol dog embraces so many unlike individuals of diverse size and form; it bothered him that the dog at three fourteen (seen from the side) should have the same name as the dog at three fifteen (seen from the front). His own face in the mirror, his own hands, surprised him every time he saw them.[1]

••• In his 1942 short story, "Funes the Memorious" (*Funes el memorioso*), Jorge Luis Borges explores the limits of memory. After falling off his horse, the story's protagonist, Ireneo Funes, wakes up to a completely changed reality. Physically, he is crippled and damaged by the fall, but mentally, he is—as he sees it—enlightened.[2] After the fall, he can remember every episode of which he has been a part, everything he has experienced, and every thought he has had in relation to it. He can track the number of stars in the sky and recite every book he has read. However, with this enlightenment, he has lost a very important ability: he can no longer omit, abstract, or generalize, but instead loses himself in the contiguity of detail. Funes has not lost his ability to use language, but he can no longer differentiate among things, no longer distill or separate. To use Ferdinand de Saussure's terminology, you could argue that Funes is tied to the horizontal, syntagmatic axis, because he only "lives on" sequences and enumerations, rather than vertical, associative connections, selections, combinations, and eliminations.[3]

Borges's short story about Funes addresses a fundamental tension that I explore in this chapter: between the cumulative contiguity of detail, on the one hand, and on the other hand, the systematic selection, combination, aggregation, and—crucially—the interplay between the two, as we encounter it in data analysis, historically and recently.[4] Thus, the general idea studied in this chapter is the mode of aggregation. I explore a handful of the many ways in which data are selected, combined, and aggregated to make sense. In this chapter, I argue that employing the idea of the combination as a prism can help us understand how the aesthetic processes inherent in presentations of data—graphs, diagrammatic summaries, aggregates—not only make it possible to practically use and manage data sets, but also to render them visible and comprehensible in the first place. Thus, this chapter describes how data presentations created by aggregation work as a special kind of operative image, and it discusses how this operationality relates to questions of individuality and identity.

MODES OF COMBINING

As data-driven research methods increasingly gain ground in the humanities, as well as in scientific research more broadly, it is important to understand how they work and make us think. In principle, data may be any imaginable pieces of information or matters of fact—given things or numbers, that in some manner supply us with evidence of our surroundings.[5] The methods used to process these data are diverse and cover a broad array of fields, from simple mathematical operations, through larger statistical studies, to graphical visualizations and other forms of aesthetic presentations of data. One of the most basic—and perhaps most overlooked—ways of handling data is the initial series of processing steps, where data sets are prepared for use, specifically for the study for which they were collected, and to exactly relate to the questions, interests, and hypotheses of the researchers involved. An important consideration related to most explorative processes is how the available information is represented and sorted to make it to useful in the given context. Data rarely just occur naturally in a format appropriate for analyses, summaries, or presentations, and the immediate condition of the data set is not necessarily the most suitable for the specific experiment the researcher has in mind.[6] "Data [do] not just exist," Lev Manovich explains in *The Language of New Media*, "they first need to be generated,"[7] just as—according to Lisa Gitelman and Virginia Jackson—they "need to be imagined *as* data to exist and function as such, and the imagination of data entails an interpretative base."[8] " 'Raw data' is an oxymoron," Gitelman and Jackson insist in their anthology of the same name, referring to Geoffrey C. Bowker's statement that "despite the ubiquity of the phrase *raw data*—over seventeen million hits on Google as of this writing—we think a few moments of reflection will be enough to see its self-contradiction."[9] Data are always already "cooked" and never completely "raw":

> We need to open a discourse—where there is no effective discourse now—about the varying temporalities, spatialities and materialities that we might represent in our databases, with a view to designing for maximum flexibility and allowing as possible for an emergent polyphony and polychrony. Raw data is both an oxymoron and a bad idea; to the contrary, data should be cooked with care.[10]

Various forms of mathematical transformation should not only be seen as normal practice when it comes to data-driven research[11] but according to Bowker, should also be considered a necessary first step to understand and use collected data, and the first step toward a discourse that includes the cooking process itself, so to speak. Cooking data is a remarkably widespread, effective, and often necessary procedure, whether it is achieved through simple mathematical transformations such as reciprocal values, logarithms, cube roots, square roots, exponentiation, or more complicated mathematical rephrasing.

The second digital mode of operation addressed in this book comprises associative and aggregative transformations such as those just discussed. Initially, "transformation" may be understood as the specific statistical, data-science meaning of the word, for example, the mathematical function T, which assigns every value of x in a data set a new value, $T(x)$, or a function that combines several values into a new aggregate, $T(x_1, \ldots, x_n)$, both "rephrasing" the original values in more suitable terms.[12] Throughout this chapter, I continuously work with the idea of data transformation as a mode of operation that defines the particular way in which you think when you undertake data-driven research. Through a close reading of Adam Broomberg and Oliver Chanarin's project, *Spirit Is a Bone* (2014), I examine the limits of the human face, understood and approached as data: What are the *effects* of data transformation, and how can an artistic intervention such as Broomberg and Chanarin's help to expand and nuance our understanding of this particular mode of operation?

In the *Spirit Is a Bone* project, the aggregative, or combining, mode of operation is central to data transformation and is examined closely by the artists. In the project's individual pieces, the two artists reshape photographic material from CCTV surveillance cameras into screen-based, three-dimensional[13] models, using advanced software that can erase all visible, personal features of the faces portrayed (e.g., facial expression, gesture, personal artifacts as jewelry, scarfs, glasses, and makeup). The outcome is a series of strange-looking digital portraits: subjects whose gaze never entirely connects with either the photographer or the viewer, instead retaining a sense of distant detachment. The project thematizes the boundaries of what a face is or can be and questions the parameters of individuality and identity. At what point does a face lose its familiar traits? Which attributes are the most important to defining a face, and how do we differentiate between what is noise and what constitutes the core essentials of a face?

In this chapter, I focus on one particular form of data transformation, namely aggregation, itself an umbrella term used in the field of statistics. Aggregation is not just one of the oldest forms of data transformation, it is also one of the most radical. In the nineteenth century, the term *aggregation* denoted the "combination of observations," a definition that conveys the idea that there is a gain in information to be had from summarizing them, beyond what the individual values of a data set tell us.[14] In this understanding, the statistical summary covers more than its parts. The sample mean is the example of this mode of operation that received the earliest technical focus, but as a theoretical way of thinking, it includes other summary presentations, including weighted means and the method of least squares, which, in essence, is a weighted or adjusted average purged of the other characteristics of individual data values.[15]

An early example that presents the paradoxical nature of the statistical summary comes from sixteenth-century Heidelberg, in Germany, where mathematician and *Stadtschreiber* Jacob Köbel worked on determining the average, but also the "right and lawful"—that is, officially approved—unit of measurement of the rod or *rute*.[16] At this time, land was measured with a unit that was about sixteen feet long. However, the question was, *whose* feet: as a rule, the King's feet were the most distinguished—and therefore more right than a peasant's. However, it was very impractical to redraw maps and renegotiate land contracts every time a new monarch was crowned. Thus, Köbel resourcefully suggested measuring sixteen representative citizens' feet, adding up the measurements, and calculating an official and general rod (as illustrated in figure 5). The participating citizens would of course be carefully se-

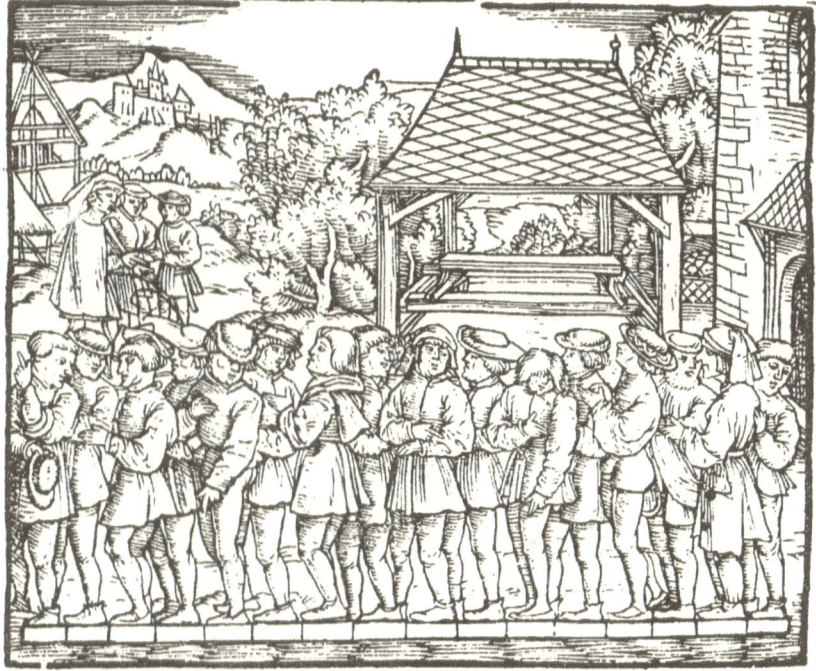

FIGURE 5. Jacob Köbel's (1460–1533) drawing of the determination of the "right and lawful" unit of the rod. Sixteen respectable citizens are lined up toe to heel, after which the measurements of their feet are summed, and this sum is divided by sixteen. Copperplate engraving from the 1535 edition of Köbel's *Geometrei*.

lected, having a reputation as respectable and acknowledged contributors to society—which would also help to legitimize the new unit—and therefore, the exclusive group ended up consisting of men who had just been to church. Köbel lined up the sixteen individuals, toe to heel, and measured the combined length of their feet. Afterwards, he divided the rod into sixteen parts of equal size, each representing one foot. We could say that the differences among the individual foot lengths were averaged out in the aggregate, yet, as Stephen Stigler points out, the idea that there was something like an average foot length, from which the unique characteristics of any particular foot were discarded, was still a long way from being recognized. The individuals whose feet made up the rod are drawn in meticulous detail in an engraving that depicts the process: it is clearly important that their identities were not discarded, as they were indeed the key to the legitimacy of the rod. The respectable and properly pious citizens walking around the city of Heidelberg indexically represented and symbolically consolidated the solid status of the average.

The taking of a mean of any sort is a rather fundamental step in an analysis. In doing this, the statistician (or computer scientist) is discarding information in the data set to make room for generalizing, condensing, or summarizing models, where the individuality attached to single data points, including the order in which they were observed, the conditions related to the observations, and the identity of the observer, is eradicated.[17] In modern data processing, you do not select particularly respectable citizens, whose individual legitimacy is believed to "stick" to the standardized unit afterwards—on the contrary, the context of the measurement is often reduced, specifically to strengthen the data's validity, as I will show later in this chapter. The reduction of context (of metadata, various forms of contamination of the signal and individuality, and deviations in the measurements) in favor of an aggregative or associative totality or set is the focus of this chapter, because it is a fundamental mode of operation in data analyses.

L'HOMME MOYEN

An example may help to elaborate and contextualize the focus on aggregation. In the beginning of the nineteenth century, discussions about the nature of the statistical summary were heating up, initially only in the natural sciences, particularly the field of astronomy, but from the 1830s on, they spread to broader cultural and societal contexts.[18] In his book, *The Politics of Large Numbers* (*La Politique des grands nombres*), statistician and sociologist Alain Desrosières describes the animated arguments about aggregation and averages that one met during the first half of the nineteenth century:

> [T]hese debates concerned the nature of the new object resulting from such calculation [of averages], and also the possibility of endowing this object with an autonomous existence in relation to the individual elements. The debate surrounding the idea of the *average man* was especially animated. It introduced an entirely new apparatus, including the "binomial law" of the distribution of errors in measurement, to deal with a very old philosophical question: that of the realism of aggregates of things or people.[19]

In referring to the binomial law, Desrosières indicates those kinds of mathematical "objects," as he calls them, related to the probability of uncertainties in measurements *not yet carried out*. In this context, the concept of the average is intertwined with other statistical phenomena that do not exist in the usual sense of the word. This is either because they rely on potential, future measurements or because they comprise a summary of observations and consequently lack a tangible presence "out there."

The statistician Adolphe Quetelet is a particularly controversial persona in the abovementioned debates. His infamous concept of *l'homme moyen*, the average human being—or, rather, average *man*[20]— was created to compare demographic groups, for example, across national borders or over longer periods of time. For instance, the height of the average English man could be compared to the corresponding French man's height, or the average height for a given age group could be followed over time to generate a growth curve for this particular population. Quetelet tried to define a model that represented each of the groups he studied, such as "criminals," "drunkards," "soldiers," or "dead people."[21] Furthermore, he speculated about the particular characteristics and nature of these groups, and expressed his unreserved optimism regarding the average, in his 1835 book, *A Treatise on Man and the Development of His Faculties* (*Sur l'homme et le développement de ses facultés, ou, essai de physique sociale*): "[A]n individual who should comprise in himself (in his own person), at a given period, all the qualities of the average man, would at the same time represent all which is grand, beautiful, and excellent."[22] However, Quetelet's ideas about demographic "averages" was met by the harsh criticism of several colleagues. The French mathematician and economist Antoine-Augustin Cournot, who at the time was considered an authority on the field of probability theory,[23] was particularly repelled by Quetelet's project. In the introduction to his book, *Exposition de la théorie des chances et des probabilités* (*Exposition of the Theory of Chances and Probabilities*), of 1843, he polemically proclaimed that he would "bring the rules of the calculus of probabilities to those who have not cultivated higher mathematics."[24] An average of all sides of a bunch of right triangles, he held, would not be a right triangle, and a totally average man would not be a man but some kind of monstrosity.[25] In 1865, another harsh critic, the doctor Claude Bernaud, wrote the following about Quetelet's average:

> Another frequent application of mathematics to biology is the use of averages which, in medicine and physiology, leads, so to speak, necessarily to error . . . If we collect a man's urine during 24 hours and mix all his urine to analyze the average, we get an analysis of urine that simply does not exist; for urine, when fasting, is different from urine during digestion. A startling instance of this kind was invented by a physiologist who took urine from a railroad station urinal where people of all nations passed, and who believed he could thus present an analysis of average European urine![26]

Quetelet, however, was undeterred by criticism. He was convinced that he could identify an average man for every age, occupation, location, or ethnicity. Moreover, he claimed that he could find a method to predict why a given individual belonged to one group rather than another. Nobody had previously used statistical calculations to separate cause and effect, and Quetelet's proclamation that "effects are proportional to causes" and that "*[t]he greater the number of individuals observed, the more do peculiarities, whether physical or moral, become effaced, and allow the general facts to predominate, by which society exists and is preserved*" (his italics) was indeed controversial.[27] He even suggested that intellectual topics (politics and social science, among other things) should be studied in the same way as physical ones:

> I attribute a great importance to this observation, that "Human society, considered on a large scale, exhibits laws similar to those which regulate the material world"; that the greater the number of observed individuals may be, the more will disappear all bodily and intellectual peculiarities; and the series of general phenomena, by means of which society erects and maintains itself, predominates with remarkable regularity in their recurrence ... what in future will be wanting to us are, not methods of observation, but *observations* made in sufficient number and with sufficient care to claim full confidence for the deduced results.[28]

Quetelet insisted that the average man could act as a "typical" example of his group and thus be used in every comparative study in which other, similar "types" were included. He believed that deductive studies of large data sets ("not methods of observation, but methods of *observations*") were the future and that—if you had enough data—you could fully trust the results of your deduction. Cournot and Bernard believed that Quetelet was on shaky ground with his broad generalizations and fascination with the Gauss distribution, but he was nonetheless a very influential researcher of his time[29] and helped to pave the way for a data-driven way of thinking, where independent human phenomena were hypothesized to behave with the certainty of astronomical phenomena. Connecting human behavior to astronomical causality—to universal constants—any deviations from these rules were "deduced as perturbations to naturalized events," as Ramon Amaro states. In other words, Quetelet spearheaded a way of thinking that dangerously aligned "contingent and chaotic phenomena with a statistical order."[30]

One of Quetelet's successors was the English statistician and polymath Francis Galton. He was particularly interested in the idea that data have the inherent capacity to conform to a regular curve, where the highest point marks the *mode*. Influenced by Quetelet, he imagined a "curious theoretical law" that would be able to explain the deviation from an average and went as far as to claim that to be studied collectively, data *had to* present a distribution that fitted under the curve. In the 1870s, he went one step further in his studies and began to explore nonquantitative data. He aspired to define a criminal type, using his so-called *composite pictures*, where he superimposed photographs of people convicted of crimes to erase individual characteristics, and thereby achieve some kind of visual average or "criminal type," as seen in figure 6.[31] He searched for common features between groups of siblings, members of large families, and people suffering from the same illnesses, and even looked for visual similarities in composites of medals that portrayed Alexander the Great, in the hope of approaching a more "realistic" representation.[32]

Today, aggregating photographs of people convicted of crimes to trace underlying visual hints at the source of their criminality, for example, seems blatantly problematic, but in Galton's time there was a great interest in exploring the possibilities and limitations of combination. However, Galton also cautioned against combining things that were too *unlike*, again noting data's tendency to cluster around a common center: "[N]o more should we attempt to compose generic portraits out of heterogeneous elements, for if we do so the result is monstrous and meaningless," he firmly states, continuing,

> It might be expected that when many different portraits are fused into a single one, the result would be a mere smudge. Such, however, is by no means the case, under the conditions just laid down, of a great prevalence of the mediocre characteristics over the extreme ones. There are then so many traits in common, to combine and to reinforce one another, that they prevail to the exclusion of the rest. All that is common remains, all that is individual tends to disappear.[33]

In much the same way as Quetelet's idea of a type lies at the basis of Galton's statistical views, he also mused on the "mediocre" and "common,"

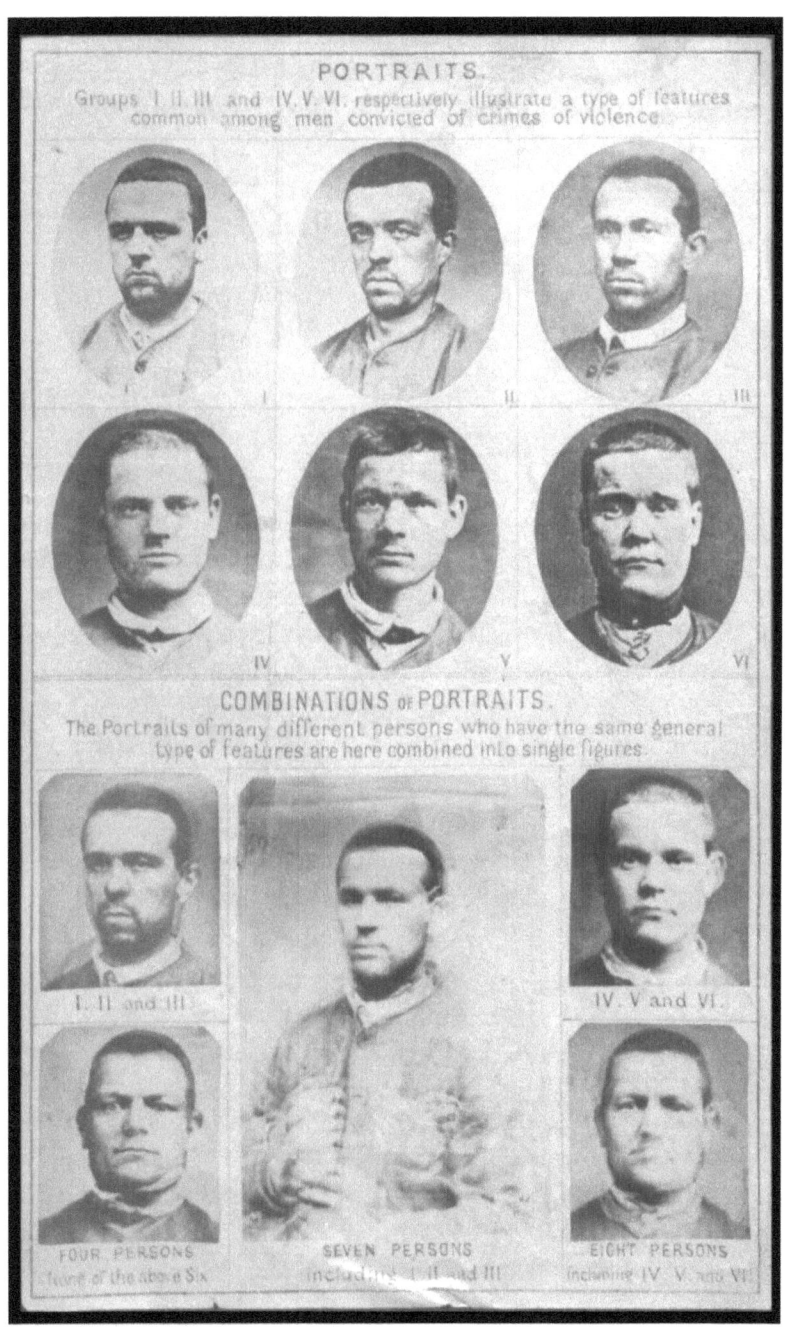

FIGURE 6. Sir Francis Galton (1822–1911), "Composite Portraits of Criminal Types" (1877). © University of London, Galton Papers at UCL Library Services, Special Collections.

as opposed to the extreme. What interested him were not the social or medical pathologies of the individual, but the generic patterns that emerged from the composites, that is, the possibilities that emerge when the individual "tends to disappear."

MODERN AGGREGATION

The animated nineteenth-century debates about generic types and average human beings testify to the conflicting understandings of data collection and processing in the scientific environments where knowledge is developed.[34] The often naturalized data transformations undertaken with respect to a data set today—for example, when calculating a sample mean—were indeed controversial just 150 years ago, and caused scientific and personal disputes. Since the nineteenth century discussions of the potential and perils of statistical transformation, aggregation as an approach to data has spread into disciplines unrelated to the natural sciences. Today, aggregated data lay the groundwork for political decisions, research into disease and public health more broadly, and for investment strategies and derivative tools in the financial sector. Aggregated data inform not only about what we know regarding the extent of the universe, they may also indicate what will happen when epidemics break out or how climate change will transform landscapes in the years to come. Data show us not only how the world looks, but also where we are heading:[35] "The data set speaks for itself"[36] and "the more data the better" seem to have been the slogans[37]—as long as we are able to handle the large (accumulated) amounts of data.[38] Data are seldomly processed individually, based on the singular datum, but are always met in sets, clusters, and masses. To find the connections and patterns, you need enough data: large data sets, Big Data, and not just single measurements or observations. Statistical data once consisted of single pieces of information that were assumed to expose real aspects of the world, but were simultaneously always contestable, because of limited realities and individual truths, Antoinette Rouvroy and Thomas Berns note.[39] Today, algorithmic processes create *dividuals*, to apply Gilles Deleuze's usage of the term,[40] rather than capture individuals; dividuals then generate countless data points that correspond to tastes, desires, behaviors, and affects, instead of self-consistent individuals.[41] Today, the idea of the data *set*—the data cluster as one, massive object, as one large, undefined mass, rather than a range of discontinuous units in a plot or a graph—prevails.

Confusion about the grammatical status of the word *data* is apparent in Google searches on this topic. Sentences that begin with "data is . . ." are now almost four times as common as sentences that begin with "data are . . . ," regardless of the extensive effort put into correcting precisely this mistake.[42] The word *data* has become a so-called mass noun, so it can take a singular verb, a circumstance that testifies to the strange position of data today: between the singular and the plural, and between being an undefined mass that spills into the real world around us and a much "harder" object of digital knowledge development. Gitelman and Jackson make a similar argument in *"Raw Data" Is an Oxymoron*, in which they describe how this development indicates the complex functions and meanings of aggregation today:

> If a central philosophical paradox of the Enlightenment was the relation between the particular and the universal, then the imagination of data marks a way of thinking in which those principles of logic are either deferred or held at bay. The singular *datum* is not the particular in relation to the universal (the elected individual in representative democracy, for example) and the plural *data* is not universal, not generalizable from the singular; it is an aggregation.[43]

In connection with the argument posed above, you could argue that the special meaning of aggregation is linked to its status as a relational operation. As an approach, aggregation is based on potential relations and emphasizes networks over hierarchies. However, this does not imply that data processing disregards hierarchies entirely. Instead, the discontinuous nature of data distinguishes them from the more familiar concept of information. Each individual datum remains distinct and detachable from the bulk, even though its representational structure aligns with other units in the collection (a point underscored in chapter 3). This trait grants data their uniqueness, enabling statisticians and computer scientists to classify, divide, embed, rank—and aggregate. However, this attribute also complicates the replication of original measurements, once they have been subjected to statistical transformation. In contrast to Enlightenment-era discussions of the relationship between the whole and its parts, and between the specific and the universal, this perspective is not as relevant to the question of data versus data sets and single measurements versus Big Data. Gitelman and Jackson emphasize that the singular datum is not particularly related to the universal, and the data cluster is not universal, not generalizable from the singular. It was precisely this aspect that concerned Quetelet's opponents when he studied the potential of the average. Should you remove information to gain knowledge—add, reduce complexity, and reject, to reach the actual signal?

You could argue that the better you know the qualities of your measurements and the context of an observation, the harder it becomes to ignore individuality in the aggregation process. At the same time, it is difficult—if not impossible—to undertake systematic, data-driven studies without combining or selecting, without transforming the input derived from the observation. In Borges's *Funes the Memorious*, we are specifically confronted with the idea of an unlimited level of detail and its many possibilities and limitations. On the one hand, there is the crippled boy's impressive awareness of the number of stars in the sky and recollection of all the elements of every book he has read, but on the other hand, there are also the problems that arise from being able to remember everything but never systematizing this information. Funes is unable to translate his phenomenal impressions of the world into conceptual knowledge, because he is unable to organize the many details he sees through aggregation and combination. He is unable to establish any consistent outline for things and must constantly experience everything anew. It is interesting to further trace the outcome of this plethora of information, met in emblematic form in the experience of Borges's protagonist. What are the consequences of this changed perception of the world? What are we able to see—and know—when we have an advanced ability to abstract and think general thoughts, forget the inherent differences between things, and instead see patterns, connections, and similarities among them? Averages, types, and signals are all statistical phenomena that pertain to the same mode of aggregation—the same digital or numerical way of thinking. Next, I will trace this particular combinationalist mode or gaze, by providing a close reading of Adam Broomberg and Oliver Chanarin's artistic project *Spirit Is a Bone* (2015) and, later, compare this to August Sander's nearly ninety-years-older archival photographic project, *Menschen des 20. Jahrhunderts* (People of the 20th Century). By comparing the two projects, I delve into the various forms of aggregation logic at play. I investigate the photographs' technical composition and stylistic arrangement, as well as the more open, interpretative elements. I explore how the two projects negotiate the aesthetics of the relationship between the individual's subjectivity and the "mass" or study population, and ultimately integrate this analysis with general questions regarding physiognomy, modern biometrics, and the implications of data-driven surveillance.

FUSED DATA MAPPING

Adam Broomberg and Oliver Chanarin's artistic project *Spirit Is a Bone* explores the sensorial effects of aggregation. Through several exhibitions (in Antwerp and Aarhus, among other places) and in a book published in 2016, the project examines a range of "three-dimensional"[44] data mappings, which represent a group of selected Moscow citizens. The images were created by a machine: CCTV surveillance images are processed by a facial recognition system developed for public security and border control surveillance. One of the noteworthy things about this system is that it is designed to create portraits without the subject's cooperation: four lenses operate jointly to generate a full-frontal image of the face, ostensibly looking directly into the camera, even if the subject herself is unaware of being photographed. The system was designed for surveillance of crowded areas, such as subway and railroad stations, stadiums, concert halls, or other public areas (or for photographing people who would normally resist being photographed), and to a large extent it produces its subjects, because no matter which direction they were looking, the face is always rendered looking forward and stripped of shadows, makeup, disguises, or even poise. The artists have deliberately applied software systems used on a larger scale in Russia, among other places, and, combined with the concrete basis of the project (the many citizens showed in the images), this project takes on a rather explicitly political character.[45]

In their project, Broomberg and Chanarin utilized and explored aggregation, which transforms photographic material into statistical models. However, they also exhibit and problematize their products, namely the compelling and peculiar three-dimensional models created by the image processing. Through glitches, abrasions, and imperfections, they showcase both the heterogeneous origins of the images—the data input, consisting of several photographs—and the statistical merging that has been employed to establish the systematized model, as, for instance, seen in figure 7.

Broomberg and Chanarin's images contain, on the one hand, traces of the original photographic material: the remnants of a long series of images captured by surveillance cameras, taken from various angles, under varying lighting conditions, and possibly over an extended

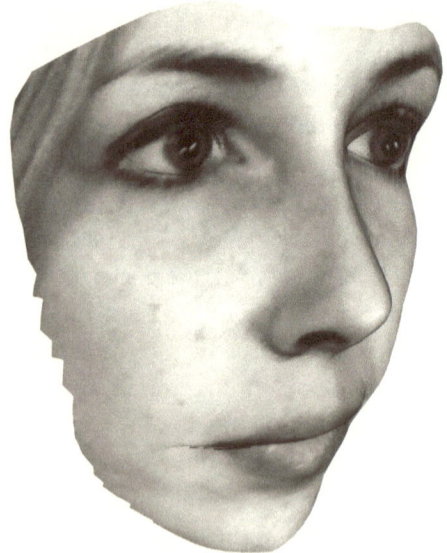

FIGURE 7. Adam Broomberg and Oliver Chanarin, "Demonstrantin," part of the *Spirit Is a Bone* project (2015). Here, a photograph in the book of the same title. © Adam Broomberg and Oliver Chanarin, MACK books.

period. On the other hand, they also narrate the story of the mechanical process that has combined them, forced them into an amalgamated whole, and given them their sculptural appearance. You could argue that they encompass their own process of creation on the visual surface, as their fused exteriors reveal the far-from-neutral aggregation process through which they were created. These images were untouched by human hands. Instead, they are the result of a series of auto-generated aggregations of disparate input, where the many small artifacts of the amalgamation remain visible, as seen, for instance, in figure 8. Why choose this distant perspective? Why employ an automated aggregation process?

Recent years have seen an important shift in the surveillance landscape, driven partly by the introduction of larger and more comprehensive data sets, partly by the extensive influence of new forms of artificial intelligence (including machine learning and automated data mining) that are able to process and make sense of these huge amounts of data.[46]

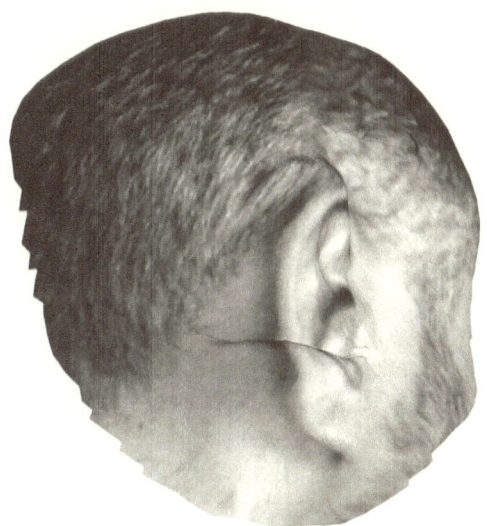

FIGURE 8. The imperfections in the skin texture are clearly visible in this three-dimensional configuration of an ear's fold. Adam Broomberg and Oliver Chanarin, "Der Chauffeur," part of the *Spirit Is a Bone* project (2015). Here, a photograph in the book of the same title. © Adam Broomberg and Oliver Chanarin, MACK books.

Both of these influences are associated with the increased automation of data collection, data analysis, and the dissemination and sharing of data sets among databases (e.g., when various public registries are interconnected for social, health, or security purposes).[47] The *SAGE Encyclopedia of Surveillance, Security, and Privacy* describes this shift as pertaining to the renegotiation of the relationship between individuals and "masses":

> Although capturing information from specific targets was the norm in conventional surveillance routines, *big data* introduces a comprehensive logic based on the premise of collecting as much data as possible: the largest number of measurable variables and indicators, for the largest number of individuals, during the longest time possible. The self-volition of individuals in the generation of data, the shift toward an extant prediction-focused perspective (tracking and storing rather than monitoring), and the increasing automation of data collection and analysis by data mining agents and data brokers suggest deep changes in traditional surveillance practices, in which the surveillant and the surveilled had clear roles and positions in the hierarchy.[48]

Traditional surveillance has been supplemented by new technical possibilities, allowing for the analysis of larger-scale connections (patterns, correlations) than previously possible[49] and the integration of databases that were previously unconnected. The goals for these vast repositories of data are closely linked to various forms of automation. This is not only because handling the accumulating volumes of data requires new, sophisticated statistical models and computational calculations, but also, I argue, because imagination and expectations are associated with automation and the diminishing role of humans in addressing politically sensitive tasks such as surveillance. Beyond obvious motives, such as improving efficiency and cost-reduction achieved by reducing working hours dedicated to various surveillance tasks, there are indeed more fundamental questions at stake.

DISEMBODIED APPARATUSES

My first question, which I believe is relevant to an extended discussion of Broomberg and Chanarin's works—and the broader discussion of automation and aggregation that they prompt—concerns the idea that humans make more mistakes than machines, that they taint observations and decisions based on their human inclination to find intentions and purposes, and their desire to simplify and reduce, to make them more manageable. The most apparent way in which the artists negotiated the challenges associated with human intervention in data collection and processing was their use of surveillance techniques. We all recognize the ubiquitous surveillance cameras from numerous magazines, movie posters, book covers, and artworks that address the theme of Big Brother or the panopticon figure, where a distinctive grey box immortalizes every single movement.[50] This hidden eye—the panopticon or Big Brother, which painstakingly captures and stores everything it sees—is also at work in Broomberg and Chanarin's project. However, what is thematized here is not so much the symbolic, emblematic image of the camera on the wall or the eye that sees, but disembodied, distant surveillance. This project seems to revolve around the apparatus or technical device that does not immediately belong to any photographing person, but just "hangs there" in the streets, persistent in its documenting operationality. You could argue that the camera in Broomberg and Chanarin's works may be perceived as a technique rather than a symbol, as it enables the disembodied, nonsubjective generation of a rich image archive for every single person who passes by, and offers itself as a portrayer of the world as it is. Here, no imprecise measurements or human errors hinder "pure," sharp images from emerging. In any case, that might be the initial interpretation.

Although surveillance cameras seem to merely capture and record the world as it is through an apparently passive, purposeless functionality, they are still built, installed, and adjusted to record passersby in a specific manner and for a particular purpose. The devices that capture photographs of passersby—on the streets in Russia and in Broomberg and Chanarin's project—were created by engineers, software developers, and data scientists who have a range of purposes they wish their products to serve. They do not create neutral equipment: instead, they design highly specialized and targeted tools that are meant to function in specific contexts. In the book associated with this project, Broomberg and Chanarin enter into dialogue with Eyal Weizman, describing their encounter with the professionals who develop modern surveillance technologies, including the Russian state apparatus, and the discursive conceptual universe in which these professionals operate:

> What first sparked our interest when speaking with these engineers, was the technical challenge they faced in producing what they call "non-collaborative portraits"—where the subject is neither consensual nor necessarily aware of the camera. These portraits, essentially three-dimensional data maps rather than photographs per se, form a digital archive that can be rotated in space on a computer screen. There is never a moment in the capturing of the "image" when human contact is registered; the subject's gaze, or any connection between photographer and sitter that we would ordinarily rely on in looking at a portrait, is a complete fiction in this space. What we're seeing is the negation of that humanity: the digital equivalent of a death mask.[51]

The developers' concept of a "non-collaborative portrait" is interesting in the context of automation in surveillance. Traditionally, when a subject poses for a photograph, there is both a photographer who frames the subject and composes the image, adjusting the photographic equipment correctly, and the person to be portrayed, who looks up, down, sideways, or away, smiles, squints, or glances back at the camera. However, in non-collaborative portraits, there is no human contact to be seen. There is no longer a photographer, no specific setting for each portrait, and the individuals depicted may not even realize that they are photographic subjects. At no point is there a human connection in the images, and a subjective gaze is never registered.

Broomberg and Chanarin even go so far as to point out that their images are in fact not images at all, but a form of visual data mapping, a series of digital death masks, or a kind of "negation of humanity."[52] In other words, one could argue that the surveillance camera behind these images objectifies the human face to the point where it ceases to be human. There is no longer any trace of contact in the eyes of those portrayed, no intentionality, no consent. The individuals depicted are presented as things to be systematically mapped and archived, ideally appearing as neutral as possible, see for instance figure 9. Thus, the cameras significantly transform the "real world out there," the data that are collected, making them accessible for the purpose of producing these images. These data are also "cooked," to use Geoffrey Bowker's expression; they are also constructed and manipulated to serve a specific purpose and function, even before they are algorithmically transformed into three-dimensional aggregates.

One of the other ways in which Broomberg and Chanarin's works negotiate the idea of humans as a source of error is through the "purifying," algorithmic postproduction they undergo as part of the aggregation process. They draw on modern face recognition technologies in their works, which aim to capture and archive individual similarities in faces. There are essentially two forms of face recognition algorithms:

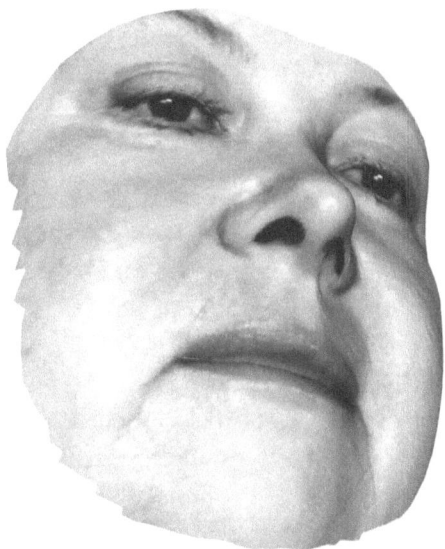

FIGURE 9. Adam Broomberg and Oliver Chanarin, "The Society Lady," from the *Spirit Is a Bone* project (2015). Notice the three-dimensional character of the model's face. © Adam Broomberg and Oliver Chanarin, MACK books.

one is related to images and the other to spatial, topographic models.[53] Although image recognition algorithms process two-dimensional images, for example, by examining matching points and distances between various parts of the face, spatially based face recognition algorithms analyze three-dimensional models. Although the first type of algorithm struggles to recognize faces subjected to various forms of camouflage,[54] the latter is more resistant to changes in the face. In this case, the images are stretched over a topographic model (or over the skull, if you will) on which the spatial contours of the face may be reconstructed with surprising precision, despite camouflaging changes on the skin's surface.

This technique involves comparing at least two photographs taken from at least two different angles, patching them together, and then creating a three-dimensional model of the face. If certain elements change—for example, if the photographed subject wears makeup, glasses, scarves, piercings, or other forms of modification, the three-dimensional recognition will still offer a reasonably robust model of the face's shape. However, this three-dimensional recognition is not unproblematic in its relationship to ideas of human error and misunderstandings. In the interview "The Bone Cannot Lie," which is included in Broomberg and Chanarin's book about their project, Eyal Weizman traces the underlying ideas built into processes of algorithmic aggregation:

> The theory is that whatever exists on the surface of the skin is seen as a potential camouflage, but that you cannot in fact change the underlying bone structure beneath the face—or not easily. So we return to the famous words of the great gravedigger and forensic anthropologist Clyde Snow: "bones make great witnesses—they never forget and they never lie." It also implies that the living face can lie: the face is a willful expression of an identity; you can smile, you can apply camouflage to it, you can fake your facial expression, whereas the assumption is that the truth is locked within the passive materiality of the bone. Snow, of course, . . . was trying to reconstruct the past, he studied lives lived and that life registered in the texture of the bone. In that sense the bones are like a pho-

tograph exposed to all influences of a life—temperature, labour conditions, illness, nutrition and so on like a negative is exposed to light. It is a slow and long exposure.[55]

The aggregation technique that combines photographs taken from various angles, in varied lighting and at different times, offers a more stable and "truthful" model of the face, according to this idea. By preventing individuals from manipulating their appearance through clothing, makeup, or facial expression, the model can reveal the "true" person in the image: the stable or basic form of the face and its underlying bone structure. Again, the aggregation may be understood as a technique with a clear purpose and obvious intention, namely, the hope of mapping what is believed to be the most "truthful" face or—to be more precise—to map it in a way that allows it to be recognized as itself as effectively as possible, even when subjected to change. Once again, a core principle of aggregation may be emphasized: by removing information, in this case data about the face's diverse expressions and texture, we may gain additional information. Measurements may be unclear, erroneous, contain deviations, or even have been subjected to various intentional manipulations. This risk is reduced when aggregation is applied, as it nullifies the potentially distorting influence of individual cases in the broader context. However, the aggregation is also precisely what removes, reduces, simplifies, and combines: the individuals' ability to express themselves—religiously, politically, culturally, or sexually—is suppressed in favor of a statistical model. In the interview, Eyal Weizman connects this new figure or way of thinking to the history of photography as a representational medium:

What I see in the archives [Broomberg and Chanarin] have created is the wrapping of the photograph, like a skin or a foil, onto an object. The result is a document that ultimately exceeds the photograph: it has become a documentary sculpture which is a three-dimensional object that is instant representation. This new type of object operates between presence and representation, and comments on the history of photography in more than one sense.[56]

Broomberg and Chanarin's works—as well as the surveillance images they resemble and artistically and politically reference—operate between presence and representation, Weizman writes, referring to the history of photography. They look more like sculptures than images, as they do not merely remain in the traditional two-dimensional realm of photographic representation but take on a three-dimensional appearance, "like skin or foil." The aggregation technique is not a form of representation that neutrally presents a face as it is, but instead is a technique that models the face into a form that promises to appear as neutral and purposeful as possible for the recognition process. Here, aggregation functions as a mode of transformation that prepares the data sets—the many two-dimensional surveillance photographs taken from various angles and in varied light—for the analyses of them that one (e.g., a Russian police officer) wishes to perform.

OPERATIVE IMAGE SCULPTURES

According to Weizman, Broomberg and Chanarin's images negotiate the status of photography as a representational medium that captures and documents the world around us. More than just photographs, they become objects of a specific way of thinking: they become models that present specific content under a set of predefined conditions. Following this line of thought, one could even take it a step further and interpret Broomberg and Chanarin's "image sculpture" in terms of the operative approach that has recently been promoted by a number of art historians and image researchers,[57] an approach that calls for a general shift away from the representationalist paradigm of images, in which they are understood primarily as documents or signatures, to a concept of images as active, "living" things:[58] in Harun Farocki's words, images, when understood in this way, "do not represent an object, but rather are part of an operation."[59] They have specific, practical purposes (in the case of Broomberg and Chanarin's works, surveillance and recognition), and they play an active or performative role in visualizing the world. Examples of contexts that have seen the increased impact of active image operations include war and airspace (e.g., smart weapons and machine-readability of landscapes, territories, and objects, including people), but also architectural modeling, traffic control systems, the construction of affectively charged environments such as malls, and other examples that have also paved the way for our current AI culture.[60] All these are image technologies that not only transmit content, but also intrude into and cocreate the contexts in which they exist. In the article, "Invisible Images (Your Pictures Are Looking at You)," artist and researcher Trevor Paglen describes this transformation of images as follows: "Images have begun to intervene in everyday life, their functions changing from representation and mediation, to activations, operations, and enforcement. Invisible images are actively watching us, poking and prodding, guiding our movements, inflicting pain and inducing pleasure."[61] It is no longer just we who contemplate images: images have begun to "look back," Paglen writes. If we apply this approach to Broomberg and Chanarin's project, we may argue that the images of faces of passersby who are looking away (or indeed not looking at all) are far from stable representations or documentation of people walking the streets of Moscow, but instead, operative aggregation models that privilege a particular idea of neutrality by virtue of the algorithmic merging used to create the three-dimensional data mapping. Thus, "[a]lgorithmically enabled image applications do not simply reproduce pre-given realities," A. S. Aurora Hoel and Frank Lindseth state, "but exercise transformative powers on both ends of the subject-object relationship."[62] Algorithmic mediation, in this case, advanced algorithmic aggregation, reconfigures the relationship between the observing subject and the observed object in a way that "ultimately determines what we see, and even how we see it."[63] In this way, the image sculptures invite an interpretation that seems to focus more on the operative characteristics of the images than on their traditional status as photographic representations. They do not merely document by means of their iconic resemblance to the people they claim to represent, but instead negotiate the epistemic status of their own coming-into-being in the algorithmic merging process.

Broomberg and Chanarin's three-dimensional data mapping could be read in terms of *technical metapictures*, a term adapted from W. J. T. Mitchell's picture theory.[64] Mitchell distinguishes metapictures from traditional pictures and emphasizes that they are indeed "deeper" than the pictures we know, as they incorporate a recursive logic: instead of just representing the object they visualize, they are representations of representation—"pictures about pictures."[65] They have an epistemological power that equips them with a sort of agency. Mitchell writes that metapictures "don't just illustrate theories of picturing and vision: they show us what vision is."[66] They are tools that guide (sometimes even replace) human judgment, and through this, they negotiate the status of pictures as such in today's techno-social culture.[67]

The nonhuman nature of the three-dimensional mapping goes even deeper than one would immediately assume. Apart from being characterized by an overall operability that enables them to transform and alter the contexts in which the images exist, these mappings are also fundamentally nonhuman in their technical construction. Rather than addressing human actors and human ways of seeing and recognizing faces, they cater exclusively to a machine gaze, that is, to processes of machine recognition. They represent the realization of a new category of images, so-called machine vision images, which may be described as mathematical abstractions operationalized by algorithms that search for their formalized qualities.[68] In the 2017 book *Nonhuman Photography*, media theorist and artist Joanna Zylinska describes the development of machine images in the following way:

> CCTV, drone media, medical body scans, and satellite imaging, photography is increasingly decoupled from human agency and human vision . . . By this I mean that these images involve the execution of technical and cultural algorithms that shape our image-making devices as well as our viewing practices. . . . All-encompassing in the workings of traffic control cameras, smart phones, and Google Earth, photography can therefore be described as a technology of life: it not only represents life but also shapes and regulates it—while also documenting or even envisioning its demise.[69]

Zylinska argues that photographs across scales—from scans of the ventricles of the heart,[70] through heat-seeking representations of refugees on the cold Mediterranean Sea,[71] to the distant, inquisitive gaze of satellite photography[72]—redefine our technologies and the ways we perceive the world. The modeled, aggregated representations of faces created by surveillance technology are just one of the many ways in which this occurs. These nonhuman portrayals are alien to human perception in terms of efficiency, temporality, and spatiality and represent an aesthetic that was never meant for human eyes. They act as complex hyperobjects, and their fundamental nonlocality impedes any local access: at the same time, it is precisely their viscosity that makes them alter our shared visual culture so significantly.[73] Broomberg and Chanarin's project specifically addresses this paradox, as it presents the aesthetic effects of machinic aggregation and offers a visual language with which we can grasp and discuss these digital modes of operation on a more general, theoretical level.

THE INDIVIDUAL IN THE ARCHIVE

The idea of creating a comprehensive, cumulative archive of an entire population's faces is not new.[74] On more closely examining the organizing principles of the book edition of Broomberg and Chanarin's project, one finds clear references to the German photographer August Sander's (1876–1964) extensive archival project, titled *Menschen des 20. Jahrhunderts* (People of the 20th Century), the first result of the famous *Antlitz der Zeit* published in 1929.[75] Thus, each portrait in Broomberg and Chanarin's book project is accompanied by a classification, such as, "Schmied," "Mitglied eines Bauernturnvereins," or "Die Stürmerin oder Revolutionärin": titles or categories that mimic the consistent structural taxonomy of Sander's project, see figure 10. In Weimar-era Germany, he created thousands of portraits of German citizens, archivally designated as "Der Philosoph," "Gymnasiast," or "Bankier," and the faces in Broomberg and Chanarin's project are inscribed in this history through the typologization principle and appropriated categories.

The reference to Sander's archival practices and typologization also suggests other less explicit, but nonetheless important, parallels between these two photographic projects. In Sander's works, there is a negotiation of both individuality and type: of the individual's status and the category in which they are placed. In Broomberg and Chanarin's works, this theme is revived through the aggregation process. Whereas

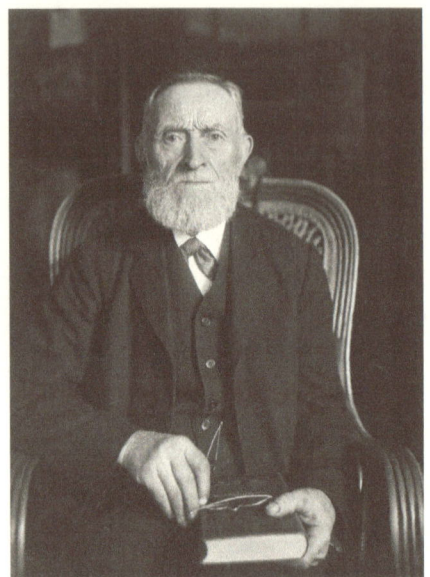 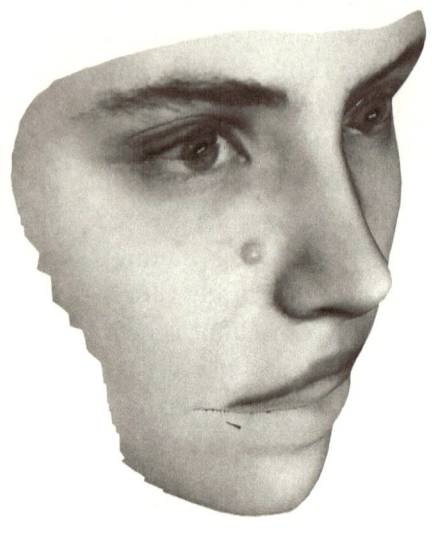

FIGURE 10. Left: August Sander, "Der Stürmer oder Revolutionär" (1925) © Die Photographische Sammlung/SK Stiftung Kultur – August Sander Archiv, Cologne. Right: Adam Broomberg and Oliver Chanarin, "The Revolutionary" (2014). © Adam Broomberg and Oliver Chanarin, MACK books.

Sander's Weimar-era works address individuality and type by relating the individuality of the subject—through the specific setting, the objects they hold, the lighting, or the composition—to their class affiliation, Broomberg and Chanarin touch on this duality in a slightly different way. Here, the relationship is a technical one, where individual photographs from surveillance cameras are accumulated, organized, and then algorithmically reduced to typological aggregates, which indicate generic categories, such as "suspicious," "high-risk," or "wanted." Although Sander's images particularly emphasize the contradiction between nonreducible individuality and generic type in the aesthetic composition,[76] Broomberg and Chanarin reveal this conflict through the technical characteristics of the photographic apparatus. A formal relationship seems to connect the two projects, as they both reference the distinct characteristics of photography, illustrating a paradox inherent to photography. As Eyal Weizman states: "[M]ore than anything else photography captures singularity, but that singularity once re-corded is also a manifestation of a type—of ethnic, gender, sociological, or economical nature—which is captured in the relation between your clothes, your facial expression, your facial hair, and so on."[77] However, Weizman does not uncritically embrace the contemporary focus on in-dividuality, which he believes has overshadowed the tension between type and individuality: "Today we are so committed to the idea of sin-gularity that type gets rejected," he writes, and continues:

> This becomes a straightjacket that is hard to escape, but one that we must escape . . . The name, when it is provided in the cap-tion, was a representation of a singularity that in the Weimar years pushed in the opposite direction than the designation of the type, which the modernist state machinery needed in place to govern. Today the situation is obviously different—state agencies look not for groups but for individuals, deviants and "unpredict-ables." State security operates in the thresholds. Face identifica-tion exists at these thresholds, initially at the entry point of a building, but now also at state borders.[78]

Weizman's criticism is directed at modern facial recognition software, which is predicated on individualized and biologically measurable suc-cess criteria and which focuses exclusively on individuality by remov-ing all context, be it social or psychological. The data sets are cleaned, and social markers, such as clothing, tattoos, piercings, facial hair, and makeup, are removed, along with psychological or mood-based mark-ers, such as facial expressions. This approach differs dramatically from the Weimar context of the 1920s, where the emphasis was on group af-filiation, rather than on the individual's physiognomy, and where the expressive capabilities of the face and the social context of the subjects portrayed were considered essential parameters when examining soci-ety based on the ideal of objectivity.[79]

Although the comparison between Sander's practices and our con-temporary image culture is productive (thanks to Broomberg and Chanarin's artistic juxtaposition), I do not fully agree with Weizman's conclusion. Although I agree with the historical reading of Sander's images and the interest in typology and class prevalent in the Weimar years, as is well documented in various other sources,[80] I do not believe that facial recognition images should be understood solely as individualized models. Instead, I argue that the parallels that Broomberg and Chanarin draw, with regard to Sander's extensive photographic project, provide useful insights into understanding contemporary digital surveillance practices as similarly typifying. The three-dimensional image objects presented by the two artists are not just examples of individual portraits, but of objectified, or even typified, data mapping, which showcases a society's citizens based on a set of predefined parameters: parameters that are particularly interesting when discussed in relation to Sander's practices and its Weimar context.

August Sander's work *Antlitz der Zeit* was just one of many photographic albums that portrayed the German population of that period. The numerous portraits in these books thoroughly investigate the facial characteristics of its subjects and highlight the prestige and status of physiognomy at that time.[81] The portraits contributed to the prevalent discourse about the "authentic" face and to a debate about whether there were consistent traits that could be traced back to particular individuals, social groups, or entire nations' cultures, based on individual faces. If this cultural reading was not achievable, it would often be attributed to some sort of cultural degeneration, as it was called.

Although conservative cultural critics asserted that contemporary people "wore their dull faces like masks," more progressive voices spoke with similar fervor of the desirability of "acquiring a mask-like, blank, impermeable face."[82] Many of Sander's contemporaries showed an interest in the new "pseudo-scientific objectivity" of his aesthetic.[83] For instance, author, journalist, and satirist Kurt Tucholsky enthusiastically wrote about *Antlitz der Zeit*, stating that Sander did not just photograph people but types. He photographed people as they are defined by their rank, profession, place of residence, class, and caste to an extent that superseded any differentiation between the individual and the group.[84] Author and physician Alfred Döblin, who wrote the introduction to *Antlitz der Zeit*, was also enthusiastic, emphasizing that images were capable of saying more about contemporary people than lectures and theories could.[85] In Sander's images, this coveted "typification" is achieved mainly through a consistent stylistic homogeneity, where the subjects are mostly portrayed facing the viewer or in three-quarter view, making eye contact, and surrounded by the objects and items that characterize their social position.[86] This approach allowed the relationship between the subject's social role in the group and the portrait's depiction of it to be as clear as possible. Hence, the legitimacy of the image as a kind of pseudo-scientific map is associated with its level of detail and the stylistic choices.

Broomberg and Chanarin's use of facial recognition software to re-interpret the surveillance camera's photographs results in distinctly different images compared to traditional ones. Statistical merging ensures that all unnecessary data are cut away, and anything that could identify the individual with regard to their rank, class, or occupation is reduced or removed entirely. Similarly to Sander's approach, the goal is to retain the essential elements, establishing the most direct relationship between the image and the individual portrayed. However, their approach is precisely that of not stylizing and aestheticizing the surroundings but to do the opposite: that is, to remove this context altogether, as effectively as possible. Although Eyal Weizman perceives this maneuver as an expression of the hyperindividualization of the subjects portrayed, where the reduction of "noise," manipulation, gestures, and other personality- or subjectivity-supporting artifacts condenses a kind of essentiality of the individual, I see it as an expression of the opposite. The two photographic strategies are very different—one stylizes and aestheticizes the subject's clothing, surroundings, and expression, whereas the other completely removes it—but these aesthetic choices still have related intentions.

In both Broomberg and Chanarin's and Sander's works, the face is treated as a kind of thing to be investigated and is consequently assigned a place in the archive or database that again refers to a large number of similar things or data points. These things may then be compared to each other, based on the predefined parameters: the folds and expressions of a typical worker's face (Sander) or the biometrically significant points for recognizing a suspicious or wanted individual (Broomberg and Chanarin). The staging of the portrayed person—the stylization or reduction of context—is just a tool used to further support this application. This is particularly motivated by the categories used as titles in both projects. Thus, Sander's "Der Stürmer oder Revolutionär" from 1925 may be compared to Broomberg and Chanarin's "The Revolutionary" from 2016, and a certain similarity may be attributed to the two images, despite the approximately ninety years between their production.

The faces portrayed in Broomberg and Chanarin's project contain both individual and typified elements, but I would still argue that the titles and categories of the works, the clear-cut aesthetic framing, and their status as modules of a larger archive contribute to a reading of their status as typified and objectified mapping, rather than traditional portraits. In the reception of Sander's images, there have been discussions about the specific photographic technique employed and whether its purpose was to highlight and preserve specific folds and expressions in the portrayed subjects' faces. In connecting the photographic depictions of facial details to the new conditions of modern existence, the stressful life in cities, and social groups' renegotiated identities, Sander has been described as a technical expert who created eloquent pictures of his time, that is, as a depicter of types as they were physiognomically defined and debated during that period.[87]

Today, facial folds are expressed as distortions and abrasions in the aggregated model: noisy glitches where the facial recognition software has not been able to successfully smooth out the expression and where the heterogeneous photographic material that makes up the model involuntarily resurfaces. Here, the "dull," "mask-like" face of contemporary citizens is caused by the removal of context and personality markers. Aggregation does not aim to find a core of identity in one particular face, but instead highlight the relationship between the individual and the numerous other faces in the archive: to systematize and generalize it, and to make it usable. Three-dimensional facial mapping is purely operative. Essentially, it is irrelevant whose face it represents if the person may be recognized and located later. The many glitches and incompletions indicate this purpose: the mapping does not necessarily need to be accurate—as long as the person may be recognized based on a series of biometrically significant points, the collapsed, depersonalized mask is successful.

THE AESTHETICS OF AGGREGATION

In both Broomberg and Chanarin's and Sander's artistic projects, aggregation seems to be a ubiquitous mode of knowledge development. Whether seen through Sander's meticulous considerations of image composition, facial angles, and expressions, or carefully selected props that bear witness to the portrayed subject's class and occupation, or through Broomberg and Chanarin's extensive image processing, where surveillance photographs are twisted and centered, color and contrast cleaned up, and topographic, three-dimensional models applied, aggregation is a pervasive approach. Geoffrey Bowker reminds us that perceiving images as raw data (whether they are arranged subject to as little intervention as possible, or come "directly" from surveillance cameras) is not just an oxymoron, but a "bad idea."[88] Data processing is transformative: Whether you are grouping, summing, or simply finding a mean, sorting and manipulation happen in a context, and it is this kind of manipulation that I refer to when I suggest focusing on the aesthetic levels of aggregation.

When you aggregate observations, you rely on a premise of reduction: something must be cut away or pushed into the background to allow something else to become distinct, weighed, and meaningful. From Borges, we know that generalization and abstraction are necessary to navigating the richness of existing in time and space. To perceive cross-cutting patterns, connections as they emerge over time, or to recognize a signal in a reality that may seem discontinuous and filled with noisy information, you must be able to combine and merge. The aggregation techniques I have explored in this chapter are all tied to specific practical purposes, and each in its own way actively or performatively contributes to the visualization of the world they are tasked with managing. They not only uncover or document, but also objectify, amplify, and reinforce certain narratives and ideas along the way. In image aggregation (e.g., facial recognition software), you search for a signal that may be condensed and coagulated through the aggregation of various elements. Thus, aggregation becomes a way in which a signal may be cultivated and made visible to the naked eye, and it is through this process that it becomes an aesthetic strategy, rather than merely a statistical or technical informational-theoretical tool.

As digital logic—or more precisely, as a *digital object* of a particular theoretical, algorithmic mindset—the mode of aggregation is special. It not only combines and selects given content, but also generates and manufactures meaning of its own. The average—the average foot, the average person, or the most important biometric points in a face—does not exist in the usual sense of the word but is produced with the purpose and the intention of seeing a connection that would otherwise not be visible. Similarly, the type, as we meet it in Sander's and Broomberg and Chanarin's projects, should be related to its categorical designation, to the processes by which it is condensed and amplified, and to the maneuvers by which it is constructed—reified—as a powerful technoscientific, sociopolitical, and philosophical tool. Therefore, it is crucial to systematically examine and conceptualize the aesthetic aspects of aggregation: aggregation is a not just a practical way of storing or making sense of data, it is a way of preparing—cooking—data sets, so they conform to a particular way of thinking, and thereby also acting, in the world.

Three

PATTERNS

The name of a thing is entirely external to its nature. I know nothing of a man if I merely know his name is Jacob.[1]

• • • Pseudomorphosis is a morphological resemblance between two forms that may resemble each other—or even be identical—but nonetheless are entirely unrelated with respect to their origins. Erwin Panofsky uses the concept of pseudomorphosis in his book *Tomb Sculpture* to describe the uncanny similarity between Carthaginian sarcophagi from the third century BCE and High Gothic funerary sculpture some 1,500 years younger.[2] In both cases, a human figure, apparently standing, is positioned on the ridge of a small, slanted, roof-shaped lid of what looks like a sarcophagus, a bed, or even a table, its head supported by one or two pillows.[3] Although the morphology is similar in both kinds of tombs, the historical processes that led to these two occurrences of the same shape are far from alike: the Carthaginian sarcophagi hold a depiction that was originally three-dimensional and recumbent, "precariously placed on the roof of a house-shaped sarcophagus," whereas the medieval tombs present an originally two-dimensional figure, "depicted on a slab in the pavement but represented as standing." The three-dimensional fullness was added later, "the figure expanding into a statue, the slab raised upon supporting members or growing into what is known as a *tumba*."[4] Unfortunately, Panofsky never elaborates on the conceptual significance of this uncanny similarity, and he does not expand on the potential of this idea.[5] However, the idea of a morphological similarity that does not arise from a shared origin is an interesting one to pursue in this chapter and its exploration of the role of patterns in data-driven thinking today. Morphological similarity is the starting point of many kinds of analyses, as the experience of external similarity (form) stimulates interest and hypotheses, followed by comparisons based on other, hidden or inaccessible, internal parameters, such as characteristics, context, or meaning (content). Comparison often depends on a certain level of comparability: there must be some resemblance between the two compared subjects, items, artifacts, traditions, or cultures for them to be compared with each other. There must be an initial pattern, line of connection, or similarity—or else it must be created artificially.

In this chapter, I delve into the profound philosophical, technical, and aesthetic aspects of the concept of patterns and engage in a broader discussion of the generation of knowledge through so-called data-driven approaches. Advancements in data processing, including the utilization of machine learning and other "intelligent" systems, have led to the emergence of new statistical objects, enabling novel classifications and virtual connections. Consequently, the scope of our understanding and the possibilities for comparison have expanded. To begin this exploration, I anchor this chapter in Stéphanie Solinas's art project *Dominique Lambert* (2004–2016), a project that raises numerous metatheoretical and methodological questions regarding pattern recognition, comparison, and comparability.[6] I argue that the Dominique Lambert project not only challenges us to reconsider the boundaries of comparison, but also compels us to explore the fundamental nature of patterns and their significance in humanist knowledge production. By examining how patterns and comparisons are understood and legitimized, I aim to shed light on the multifaceted dimensions of these essential aspects of scholarly inquiry.

LIVING LOOK-ALIKES

Stéphanie Solinas's book on the enormous mapping project *Dominique Lambert* (2006–2016) abounds with nested patterns, potential lines of connection, and (im)possibilities for comparison. Like pearls on a string, translations and transmissions of photographs, surveys, portraits, psychological descriptions, and simulations unfold throughout the book's hundreds of pages. "Who is Dominique Lambert?" is the project's central research question, and the answer is found in the French *Pages Blanches*: 191 Dominique Lamberts are listed in the telephone directory. Dominique is the most popular gender-neutral first name in France, and Lambert is the twenty-seventh most common last name. Together, they comprise a relatively ordinary name, statistically speaking. Taking these 191 Dominique Lamberts as her point of departure, Solinas established her study population. They each received a letter with a survey and a personality test. Sixty-five replied, and twenty responded to the request to send a passport photograph to the artist. These twenty people, all named Dominique Lambert as the essential common feature, constitute the project's empirical basis, whereas the name itself becomes the cause and starting point: the ephemeral product of the study, the general idea or obsession. Throughout the pages of the book, and in several exhibitions, the figure of Dominique Lambert is minutely scrutinized. On one page, Dominique Lambert has a pale and freckled face, on another she is a red-haired Frenchwoman with a nose piercing and a slightly thick neck, and on a third page, he is a twenty-eight-year-old loner with a middle parting. This face, that state of mind, and these qualities all seem to be inapplicable and incoherent answers to an absurd question: Who is Dominique Lambert?

In the mold constructed by Solinas's question, a stream of Dominique Lamberts pours in, solidifies, and takes shape into the form the project pursues: Dominique Lambert as an abstract category and name, but at the same time, a category comprised of data points: an agglomerate of living, individual components which, together, establish the general definition. Solinas's experiment is in many ways an absurd and parodic one and is perhaps best characterized as an artistic appropriation of a formal-mathematical vocabulary.[7] It partakes in and reflects on the pseudomorphological problem described above. Some of the Dominique Lamberts, we will soon see, come to resemble the constructions of the experiment. Images come so close to each other that a connection and a standard of comparison emerge for the inspecting, analytical gaze.

Solinas's whimsical mappings give rise to a range of methodological questions related to the relationality of comparison and comparability. What is one to do with an artistic investigation that defines such a distinct product or object—that is, Dominique Lambert—but at the same time, fundamentally breaks with the very idea of what such an object is? Or, rather, it breaks with the usual idea of objects and things as concrete, physical entities that may be perceived in time and space and have direct, causal effects. Dominique Lambert does not exist "out there," and Dominique Lambert is not a meaningful, scientific, or terminological category. Even so, Solinas has, over a decade of studies, through books and several exhibitions in France and abroad, meticulously examined Dominique Lambert as precisely that—a category or figure that is exceedingly meaningful.

In this chapter, I argue that the abovementioned art project, equally complex and curious, methodical and facetious, rigorous and flexible, offers an exceptional starting point for a metatheoretical and methodological discussion of questions related to comparison and comparability: the alluring nature of look-alikes and their influence on analysis and interpretation and the search for patterns and connections so prevalent in data-driven knowledge production. Comparison is not only related to discussions of the potential and perils of comparatism, a prominent strategy in studies of art and literature (as in comparative literature),[8] but is also discussed here in a broader intellectual framework. The *Dominique Lambert* project questions not only how and when something may be compared, but also the more general nature of comparison and its legitimacy as a research approach in the humanities. Comparison is one of the many ways in which knowledge is presented (as in both provided and shown), so the *Dominique Lambert* project concerns the more general discussion of the relations between data and information, and information and knowledge, and the techniques and methods that convey these processes of comparison in the first place.

A range of questions immediately arises when we encounter Solinas's project: What variations, distortions, and gaps emerge between the many remediations[9] (between photography, surveys, drawn portraits, psychological descriptions, and computer simulations) and the various processes of transmission and translation? How should an exploration of the *Dominique Lambert* project be positioned, or, rather, which state of the translation should be analyzed, which object should be studied? Who or what *is* Dominique Lambert? In the following pages, I discuss these questions, and trace the essential figure embedded in the layers of comparison, that is, Dominique Lambert as a working classification system and an obsessive search strategy. More specifically, the analysis takes three fundamental questions related to the treatment and representation of the project participants as its point of departure: first, the many men and women with nothing in common but their name; second, the translation between the various series of data input fed into the mapping—the surveys, personality tests, computer-simulated facial models, drawings, descriptions, and portrait photography that constitute the project's technologies and methods; and third, the central idea of a pattern or connection that, according to the project's internal logic, is shared by the many individual participants. In extension of this analysis, I intend to describe and conceptualize the ways in which this project operationalizes comparison and translation as examples of virtual, yet nonetheless effective, modes of operation.[10] At the end of this chapter, I discuss Solinas's project in relation to data-driven knowledge production and its comparative search for patterns and connections.

WHO IS DOMINIQUE LAMBERT?

> *Dominique Lambert is a forty-seven year-old man. He is finicky and meticulous. He has freckles on the face, fair skin. His hair is white, but it was red when he was younger. He is clean and shaven. His nose is turned-up, slightly.*[11]

For Stéphanie Solinas, the starting point of her art project was the name Dominique, as it is the most common French gender-neutral name. The project began when Solinas sent a letter to each Dominique Lambert, requesting that they complete what is known in France as a "Chinese portrait" (a metaphorical description in which one compares oneself to other things or elements, e.g., "If I was a color, I would be *x*," or "If I was a vehicle, I would be *y*"; examples of responses received by the artists include, "If I was a dish, I would be a rock lobster" and "If I was a monument, I would be a pyramid").[12]

For each of the Chinese portraits received (see figure 11), a written

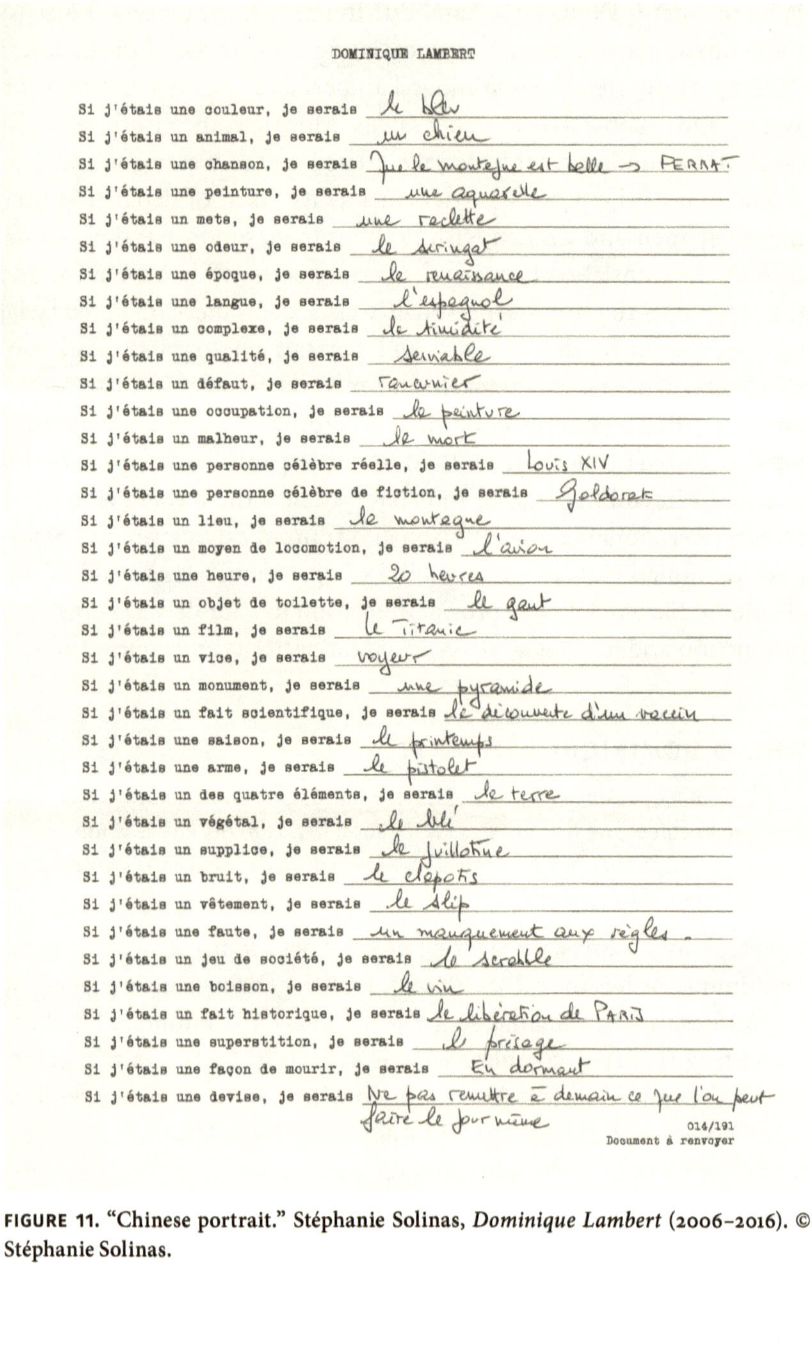

Si j'étais une couleur, je serais _le bleu_

Si j'étais un animal, je serais _un chien_

Si j'étais une chanson, je serais _Que la montagne est belle → FERRAT_

Si j'étais une peinture, je serais _une aquarelle_

Si j'étais un mets, je serais _une raclette_

Si j'étais une odeur, je serais _le seringat_

Si j'étais une époque, je serais _la renaissance_

Si j'étais une langue, je serais _l'espagnol_

Si j'étais un complexe, je serais _la timidité_

Si j'étais une qualité, je serais _Serviable_

Si j'étais un défaut, je serais _rancunier_

Si j'étais une occupation, je serais _la peinture_

Si j'étais un malheur, je serais _la mort_

Si j'étais une personne célèbre réelle, je serais _Louis XIV_

Si j'étais une personne célèbre de fiction, je serais _Goldorak_

Si j'étais un lieu, je serais _la montagne_

Si j'étais un moyen de locomotion, je serais _l'avion_

Si j'étais une heure, je serais _20 heures_

Si j'étais un objet de toilette, je serais _le gant_

Si j'étais un film, je serais _le Titanic_

Si j'étais un vice, je serais _voyeur_

Si j'étais un monument, je serais _une pyramide_

Si j'étais un fait scientifique, je serais _la découverte d'un vaccin_

Si j'étais une saison, je serais _le printemps_

Si j'étais une arme, je serais _le pistolet_

Si j'étais un des quatre éléments, je serais _la terre_

Si j'étais un végétal, je serais _le blé_

Si j'étais un supplice, je serais _la guillotine_

Si j'étais un bruit, je serais _le clapotis_

Si j'étais un vêtement, je serais _le slip_

Si j'étais une faute, je serais _un manquement aux règles_

Si j'étais un jeu de société, je serais _le scrabble_

Si j'étais une boisson, je serais _le vin_

Si j'étais un fait historique, je serais _la libération de PARIS_

Si j'étais une superstition, je serais _le présage_

Si j'étais une façon de mourir, je serais _En dormant_

Si j'étais une devise, je serais _Ne pas remettre à demain ce que l'on peut faire le jour même_

014/191
Document à renvoyer

FIGURE 11. "Chinese portrait." Stéphanie Solinas, *Dominique Lambert* (2006–2016). © Stéphanie Solinas.

portrait was developed with the assistance of the Advisory Committee for the Description of Dominique Lambert (composed of a psychologist, a statistician, a police inspector, and a lawyer).[13] This text formed the basis of the sketch drawn by the artist Benoît Bonnemaison-Fitte, which Dominique Ledée, police investigator for the Bureau of Criminal Identification, then transformed into a composite picture, using computational methods (see figure 12). Then, Solinas searched for models who resembled these police photographs and had a professional portrait photographer take their pictures (see figure 13). A sealed envelope containing the passport photograph of the Dominique Lamberts who were the authors of the Chinese Portraits closes the chain of representation (see figure 14).

The first questions I want to pose with regard to this wonderfully weird and scrupulously methodical art-research project are related to its treatment of data: Were the data managed in reasonable ways, or, more precisely, were they used meaningfully with regard to the project's over-

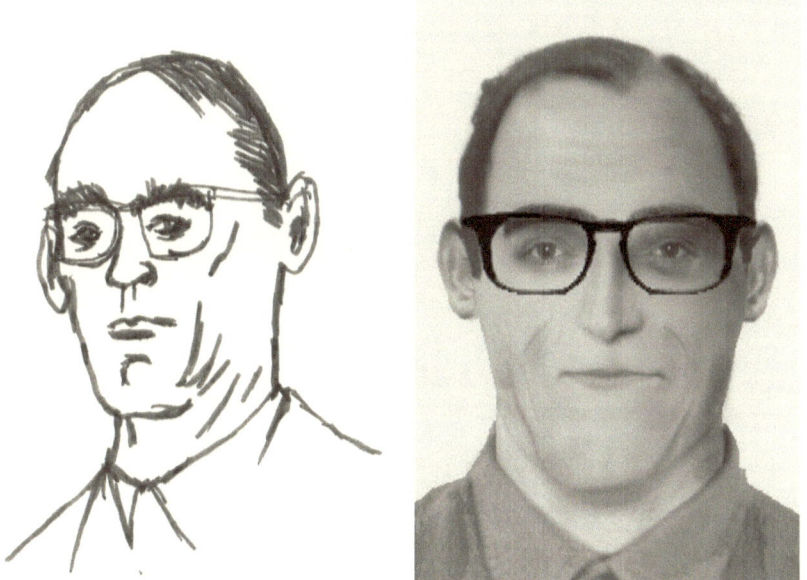

FIGURE 12. Left: pencil portrait by Benoit Bonnemaison-Fitte. Right: computer-generated portrait (right). Stéphanie Solinas, *Dominique Lambert* (2006–2016). © Stéphanie Solinas.

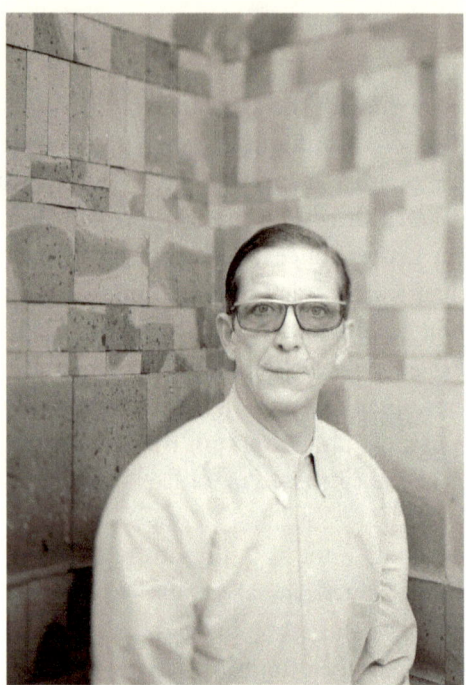

FIGURE 13. Photographic portrait of model. Stéphanie Solinas, *Dominique Lambert* (2006–2016). © Stéphanie Solinas.

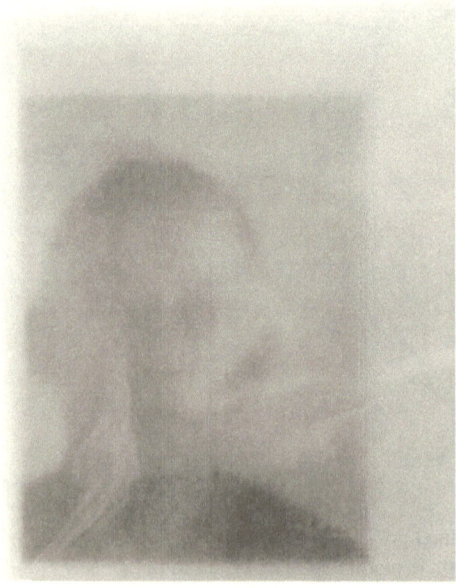

FIGURE 14. Passport photographs attached in a thin sheet, originally hidden under the cover of the book. Stéphanie Solinas, *Dominique Lambert* (2006–2016). © Stéphanie Solinas.

arching question? What is the status of all these men and women who seem to have nothing in common but their names? How were their data collected and managed, and what is the status of these data with regard to Solinas's initial research question? Each of the participating men and women—with their preferences, personalities, dispositions, and ideas—represents a multidimensional data point, an entry into the data space that is difficult to position unambiguously. When one explores the comprehensive descriptions and fragments presented in the book, this point becomes clear: comparisons between individuals, buildings, colors, dishes, languages, and toiletries entered in the Chinese portrait; descriptions of receding chins, turned-up noses, and greying temples; poorly executed computer simulations of jawlines, necks, and earrings. All these individuals were aggregated in the same data set, brought together under one category, because for some unknown reason, they were all given this name at birth. Nevertheless, the project—with its research question ("Who is Dominique Lambert?") and its obsessive idea of a general classifier—claims that this comparison is meaningful, that the "likeness" may be approximated through systematic investigation: that a kind of "shared form" exists beyond the individual data points.[14] However, the question is whether there is reason to believe that the participants resemble one another in any way; whether their proximity or contiguity within the system of classification—that is, their common name and place in the investigation—legitimizes the assumption that they are in any way similar.[15]

The *Dominique Lambert* project, with its recurring and insistent questioning, calls to mind a central question in the structuralist semiotics of Swiss linguist Ferdinand de Saussure (1857–1913), namely, how signs relate to other signs or, more precisely, the types of relations in which signs may take part. With two systematic axes, the syntagmatic axis and the paradigmatic axis, Saussure describes how some types of relations may be found within a discourse (e.g., in a text or in a conversation), whereas others are external to the discourse:

> We see that the co-ordinations formed outside discourse differ strikingly from those formed inside discourse. Those formed out-side discourse are not supported by linearity. Their seat is in

the brain; they are a part of the inner storehouse that makes up the language of each speaker. They are *associative relations.* The syntagmatic relation is *in praesentia.* It is based on two or more terms that occur in an effective series. Against this, the associative relation unites terms *in absentia* in a potential mnemonic series.[16]

Although the syntagmatic relation describes the actual, real links in a conversation, units that follow one another by virtue of the linear character of expression, the paradigmatic relation describes the substance of the association, the shared, or the recollected; in other words, the connections made by comparisons to things outside of the conversation. Although Stéphanie Solinas rather soberly examines, describes, and maps a range of possible (and impossible) characteristics of a group of selected people through the *Dominique Lambert* project, nonetheless, the investigation rests on a fundamental belief, or inherent, predetermined guiding principle. The essential claim or idea is that these people may be compared, and thus, they are not only compared, but also made comparable through their aggregation in the project. The various sequences of investigative description are connected not only by their syntagmatic contiguity—Solinas has chosen to list the Dominique Lamberts one after the other, ask them about the same things, and draw, simulate, and describe them through the same techniques—but also on the basis of the paradigm of similarity or meaningful convergence. This project both provides and reveals the connections it claims to see and thereby presents the theoretical object it studies and the investigative mode applied: as a kind of pseudoform, Dominique Lambert emerges as both a point in the data space and on an axis of continuity, and thus should be assigned both actual and virtual properties. To use Saussure's vocabulary, the comparisons made in the *Dominique Lambert* project may be described as primarily absent, potential figures in their claim to similarity, rather than in terms of actual connection or contiguity. Comparison or disambiguation through a presupposed, associative connection seems to be the project's modus operandi: Solinas persistently seems to claim that links, patterns, and meaning can be found. The project stubbornly tries to demonstrate that one may intelligibly speak of a category or object called Dominique Lambert, and emphasizes and supports this agenda through its almost pataphysically sober jargon.

The systematic principles of the *Dominique Lambert* project may be said to operate on a virtual, rather than an actual, plane: the individual participants differ widely—they have disparate, even contrasting preferences, dispositions, appearances, backgrounds, and concerns—yet, at the same time, one specific similarity makes them appear here in this study. Therefore, their name is assigned a special, predicative status in this project, because it becomes the focal point of the statement that this face, this state of mind, these qualities *are* Dominique Lambert. Pushed to its logical conclusion, the desire for coherence in this discontinuous data set is embodied by the virtual figure that the project addresses, that is, Dominique Lambert. That is the idea around which the theoretical operation revolves, that connection we imagine to be there. A figure like this does not have a physical correlate; in principle, one could change Dominique to any other name. Solinas could have studied the connection between all individuals called Stéphanie Solinas. This is the fundamental absurdity or unreasonableness of the project.

TRANSLATION BETWEEN SERIES OF DATA INPUT

A similar tendency may be found in the many sequences of Solinas's investigations: the committee of professional experts translated Chinese portraits into descriptions of each individual participant, which were then translated into a drawing, then into a computer-simulated version, and then into a photographic portrait that looks like the computer simulation, and then, finally—perhaps—the original passport photographs were hidden in a semitransparent sheet beneath the book cover. In and of themselves, all these processes seem legitimate enough, and the book even indicates that Solinas committed to selecting a range of procedures that precisely mimic (or parody) reasonable and well-founded scientific data collection methods.

She tracked down experts who could compose descriptions of the project participants, she found a visual artist who could visualize these descriptions, she searched for computational tools similar those used for criminal investigations in France, and she asked a professional portrait photographer to create portraits of models resembling the previous steps of the study. It is less the isolated series of data input than it is the combination of them that raises problems: the individual sequences may be legitimate, yet the series of sequences as a whole point out the absurdity of the project. How could a range of Chinese portraits that consist of comparisons between a participating Dominique Lambert and colors, famous fictional characters, vehicles, plants, and drinks legitimately result in descriptive sentences such as, "He looks a little disillusioned, maybe a bit cynical," "His hair is white, but it was red when he was younger," or "He is finicky and meticulous"?[17] How are these descriptions transposed into the visual realm by the artistic renditions? And, how can a computer reliably simulate the face of this "disillusioned" and perhaps "a bit cynical" person? There is a fundamental absurdity to these investigations, something impossible or even unreasonable.

The absurdity becomes a central perspective that is both prompted by and thematized through the project's systematic procedures. Solinas mimics rigorous, scientific methods, which can help to uncover the research question presented and described in soberly academic ways in exhibition texts and in the book from 2016. Before publishing what she calls the "possible faces of the 'true' Dominique Lamberts," she introduces the book in the following way:

> I established that the one hundred and ninety-one Dominique Lamberts listed in the French residential telephone directory would be my study population. I asked each of them by postal mail to complete a Chinese portrait. With assistance from the Advisory Committee for the Description of the Dominique Lamberts (comprising a psychologist, a statistician, a police inspector and an attorney), I elaborated a written portrait for each Chinese portrait returned.[18]

Through the various sequences of photographs, sketches, texts, and computer simulations, Dominique Lambert slowly emerges as a tangible object: the study population (the disparate datapoints all named Dominique Lambert) is not treated as a heterogeneous assembly, nor as a complicated, multidimensional data set or a group of living look-alikes. Instead, they are approached as a conceptual figure that may be realized in many ways. The project seems to be less focused on concrete examples of this Dominique Lambert through his or her response to this or that question; instead, the category of "Dominique Lambert" is meticulously dissected. In this regard, it seems relevant to ask how Solinas's presupposed basis for comparison is made visible in the first place: What are the prerequisites for what a statistician or a police inspector knows? How or by what means does an artist know? What is made knowable to the psychologist through personality tests or a Chinese portrait? What becomes visible in the investigation—and what remains invisible? The question about who Dominique Lambert is, is negotiated on two planes: Dominique Lambert is both visible and invisible, he is these characteristics, she is this face, he is this preference, but she is also invisible, outside of the data set, metaphorical or associative. Here, the comparisons in the Chinese portraits may be emphasized: "If I was a dish, I would be rock lobster." Something is constantly hinted at or alluded to, something absent, something outside of discourse.

THE TRUE FACE OF DOMINIQUE LAMBERT

The central fragment of the *Dominique Lambert* exhibitions and book may be the passport photographs that were sent by the participants at the beginning of the project and which now play a peculiar role in the end-product. As the single identifier—the only way for the spectator or reader to come closer to a clarification of who the "right" Dominique Lamberts are—the passport photographs are particularly puzzling. In the exhibitions, their status is unclear, as they remain inaccessible to the spectator, face downwards in closed vitrines. However, in the book, a set of empty spaces (*Leerstellen*)[19] are marked as a special section, where the careful archivist may match and paste in the right photographs in the labeled places. One could interpret this as a cute or quirky way to engage the reader or as a reference to the thorough cataloging mode of the obsolete family album.

One could also approach the passport photographs from a more "scientific" angle and enter the discussion already apparent in Solinas's introduction to her methodological approach, quoted above. From this perspective, the reader or viewer is not so much an archivist who is asked to participate in, or complete, the project, as she is an initiated coresearcher, a police inspector taking part in an important investigation, or a psychologist undertaking an important mapping of a rare personality type. One must set out to interpret the images' motifs and position them in the studies presented in the book or the displayed vitrines on the same terms as Solinas. From this perspective, the passport photographs become a kind of passkey or code, carefully kept secret: the item you *could* have, if you were in charge of knowledge production.

The foregoing gesture is particularly teasing, because it is so difficult for analysts to resist continuing the story. We cannot help but try to fit together the images and the texts, to "solve" the equation. The danger of forcing a connection, of seeing patterns where there are none, is substantial. The images are so *near* one another (contiguously), that there *should* be a connection: with all this scrupulous mapping and examination, it seems unlikely (unfair, even) that a solution would fail to appear, even though this is an absurd and unreasonable study in the first place. The passport photographs seem to be assigned this paradoxical function in the project: as scientists, we analyze, interpret, and act on available data, and when a passport photograph really resembles one of the printed drawings, simulations, or descriptions at our disposal, we tend to relate them, even though we have no certainty about this pairing. In other words, we accept that a possible similarity between a simulation, a drawing, or a description is good *enough*—that the similarity is strong *enough*—when we treat it as an actual, empirical object. However, the challenge is that these lines of connection extend beyond the data set, beyond the individuals studied in this project. In this way, Solinas's project may be the point of departure for a broader, or more general, discussion of the relations between contiguity and distance—identity and difference—when one compares subject matter. The validity of comparisons depends on assumptions, whether they stem from domain knowledge (that of the police inspector, the psychologist, or the statistician in Solinas's investigations), theory, or guesswork. One could argue that comparisons generate virtual connections because they are created on the basis of knowledge that exists outside of the data set itself: they are something we *do* with the data set, rather than something we discover within it.

COMPARISON AND VIRTUALITY

The mode of comparison as potential and problem is one of the central themes in Solinas's project. Whether it is in the comparison between two of the most important faces in the project—the original face on the passport photograph and the final output of the investigation, the professional portrait photograph—or the comparison between information at various stages of the translation process—for example, between the Chinese portrait and the computer simulation—questions of research methodology inevitably arise. Because after all, what may be compared, and is it legitimate to compare material that is so dissimilar? What kind of relational structure may be found between all these discontinuous data points (this face, that state of mind, these qualities), and can it even be measured or studied in meaningful ways?

To come closer to a more comprehensive understanding of the ways in which comparison is understood in the *Dominique Lambert* project, I draw on Gilles Deleuze's concept of virtuality.[20] This is particularly relevant here, because it helps to explain how structures or systems (whether they are social, linguistic, biological, ecological, or something else) can work without having a concrete or material form as something tangible.[21] In *Bergsonism,* memory and virtuality are linked together, as the virtual is described as something that "acts without being-present": "We have thus confused Being with being-present. Nevertheless, the present *is not*; rather, it is pure becoming, always outside itself. It *is* not, but it acts. Its proper element is not being but the active or the useful."[22] According to Deleuze, the virtual may best be described as a relation or a model that includes the processes taking place, but that cannot be understood based on actual effects. Thus, Deleuze's concept of virtuality is a way to understand that which may not be materially visible yet acts as a real possibility. The virtual is a form of potential that may be actualized; it is not a material entity, but it still works.[23] By operationalizing Deleuze's concept of virtuality, we can come closer to an understanding of how the perceived proximity (contiguity) of the various data points in Solinas's project may give rise to a sense of similarity. The metaphorical function of comparison becomes a virtual structure that operates and intervenes in the data set when it is actualized. As a concept, the virtual helps to explain how incoherent data points—men and women who have nothing in common but their name—can belong to the same relational structure or system. Conceptually, it describes how a figure or category, which does not reside "out there" and which cannot be directly observed or measured, nevertheless works and transforms the connections of which it is a part. "Dominique Lambert" operates as a category and becomes a virtual structure—a technique if you will—that forms the project's output. However, Dominique Lambert continues to be an absurd or unreasonable figure, which constantly threatens to undermine the project's legitimacy. It is precisely this threat that is interesting to trace further.

COMPARISON AND COMPARABILITY

> Much of the power and interest of many a good metaphor derives from how massively and conspicuously different its two subject matters are, to the point where metaphor is sometimes defined by those with no pretensions to originality as "a comparison of two unlike things."[24]

Stéphanie Solinas employs absurdity as one of her main tools when she investigates the potential and problems of seeking connections and patterns where there are none to find. Then, her general claim when it comes to this project is that one may reasonably discuss a category or an object called Dominique Lambert, which combines the many different features of the individual participants. The potential of this strategy is compelling, and in the last two decades, many scholars seem to have be drawn to regularities, patterns, connections, and other forms of induction in the hope of extracting emerging patterns of information from data.[25] You correlate and compare as never before, and the possibility of comparing on a previously unprecedented scales has attracted a lot of interest, not just in literary studies departments, but also in fields such as art history, musicology, performance studies, and media studies.[26]

A professionalized infrastructure of digital humanities research is easily noticed, and many conferences, academic societies, and books, journals, and articles testify to the impact of this field.[27] Computational pattern recognition is of particular interest in this context, because it is ascribed the capacity to reveal new knowledge based on data, knowledge that was previously inaccessible, either because it was out of reach for the researcher (in the "unknown unknowns"[28]) or because it simply remained invisible to the human eye.[29] In this context, scholars such as Franco Moretti, Matthew Jockers, Matthew Wilkens, and Andrew Piper have championed sophisticated machine techniques like topic modeling and network analysis, advocating for their ability to elucidate macroscale patterns of language and form distilled from extensive literary corpora.[30] Historian Chiel van den Akker writes that computer-aided pattern recognition may "allow scholars to answer new questions, widen the scale of their research and the scope of their generalizations, be more rigorous and systematic about their approach, and be more nuanced and precise about their interpretations," and notes the vast potential of pattern recognition, also within the humanities.[31] With new automatic systems—such as machine learning and artificial intelligence—a new object of knowledge emerges: new classifications become possible, and new virtual connections may be created. You could argue that techniques such as these are popular not just because of their practical usability; they also facilitate a compellingly powerful way of thinking and reasoning. Along these lines, Irish geographer Rob Kitchin has argued that an "empiricist epistemology" is finding its way into the sciences, an epistemology that seems to rule out deductive or theory-based scholarship. He argues that a new mode of science is being created in the wake of new data-driven approaches to research, one in which the modus operandi is entirely inductive in nature. Taking its point of view in large data sets, Big Data promises to deliver previously unseen patterns and insights. In Kitchin's words:

- Big Data can capture a whole domain and provide full resolution;

- there is no need for a priori theory, models or hypotheses;

- through the application of agnostic data analytics the data can speak for themselves free of human bias or framing, and any patterns and relationships within Big Data are inherently meaningful and truthful;

■ meaning transcends context or domain-specific knowledge, thus can be interpreted by anyone who can decode a statistic or data visualization.[32]

Intrinsic patterns and relations in data sets may now be detected and made visible, data can "speak for themselves," and new insights may be discovered: insights that are not only "inherently meaningful and truthful," but that transcend the contexts and domains in which they originate. This new "empiricist epistemology" identified by Kitchin is not only characterized by an increased confidence in empirical data, but also by a hostility to theory,[33] which crowns induction as an unmatched provider of emerging knowledge. At least, that is the discouraging diagnosis offered by Kitchin.

To return to Stéphanie Solinas's artistic project, it becomes clear that it provides a timely and complex position from which to open the general discussion of the chapter. In particular, it is interesting to investigate the tension between its living individual data points (the participating men and women, with their individual quirks, preferences, personalities, and imagination) and its form (Dominique Lambert as a virtual category). Where are these patterns and relationships located, and how can the nagging and compelling correlations between the original passport photographs and the engineered descriptions, drawings, and simulations be understood, if not as patterns or connections born of the data set itself?[34] What status is given to the individual thing when it meets the comparative mode of operation with its aggregation based on form?

Several challenges emerge when the individual data points—the individual persons or personal features—risk being disturbed or erased altogether in favor of a study's broader tendency or general directions. In the worst-case scenario, individuality is dissolved for the benefit of the pattern, because to make sense, the presumed similarity in the comparison presupposes a general homogeneity. You could argue that when things are made comparable, they appear to be more alike than they perhaps are. On the other hand, there is also something to gain when two unlike subjects join: new knowledge may emerge, unseen patterns may appear, and one may leave possible biases and other external framings out of account. That would be the hope, anyway.[35]

As mentioned, Franco Moretti is one of the pioneers of the use of computational pattern recognition in the humanities, more specifically, in literary criticism. One finds in his work precisely this tendency to privilege inductive methods. In his 2005 book, *Graphs, Maps, Trees*, he describes his distant reading models as having the capacity to track down what he calls "emerging qualities," an idea not far from the one we meet, in a parodied form, in Solinas's project:

> [Y]ou *reduce* the text to a few elements, and *abstract* them from the narrative flow, and construct a new, *artificial* object like the maps that I have been discussing. And with a little luck, these maps will be *more than the sum of their parts*: they will possess "emerging" qualities, which were not visible at the lower level.[36]

The artificial object described by Moretti is an emerging category that is not apparent in the literary text itself, but which takes form through the process of comparison. Like the emerging figure of Dominique Lambert, Moretti's artificial mapping also depends on the reduction, abstraction, and construction of a literary text. And like Dominique Lambert, which (or who) does not have a physical correlate but is produced conceptually, Moretti also constructs the patterns he finds in a text.

In a similar vein—thinking inductively about literature and humanistic analysis—is Andrew Piper's discussion of generality in his 2020 book *Enumerations: Data and Literary Study*. At the outset of the book, he criticizes traditional literary criticism for its inadequacy to make generalizing observations, referring to this deficiency to connect individual parts to larger wholes as an "epistemological tragedy."[37] While Moretti's distant reading primarily focuses on patterns as emergent qualities, Piper's conception of patterns instead revolves around the context critically asserted to exist, when individual works are taken as expressions of general trends: when single passages stand in for an entire "great work," which, in turn, stand in for "all of occidental literature." Instead of making claims on behalf of single works, Piper advocates for a "science of generalization," which utilizes computational models to achieve more robust overviews.[38] While this inclination to making broad assertions might seem close to Moretti's distant reading of the "great unread,"[39] Piper diverges from Moretti in his focus on representativeness (rather than mere scale) and in his advocating for the construction of more complex models (e.g., vector space models, topic models, and predictive models), which can help the critic argue more rigorously—and, potentially, also more transparently.

Moretti and Piper represent an academic field that has seen significant growth in recent decades but has also faced considerable criticism.[40] Some argue that the core of literary criticism resides not in generalization, but rather in a New Historicist approach emphasizing particularities. Others argue that, whether literary criticism should generalize or not, to do so using computational methods is not worth the political cost—for critics like Daniel Allington, Sarah Brouillette, David Golumbia, for instance, distant reading capitulates too much to the academy's neoliberalization.[41] What is important to stress here, however, is not that systematic studies of larger text corpora are problematic in themselves (or new for that matter),[42] but instead that supplementing close readings of individual works with systematic, statistical models, to see broader connections and a greater overview of the general patterns and themes, comes with certain ideas. Rather than simply mining for already existing qualities—as seems to be the dominant narrative—connections are rather created theoretically. Comparison as a methodological or technical tool for seeing something that was previously invisible to the human eye is a complex maneuver that offers both potential and challenges. To compare two subjects, you must first ensure that they are comparable, and this maneuver calls for both the correct tools and their proper use.[43] It is a way of highlighting the similarity between two subjects not particularly alike at the expense of the potential distance between them. "Much of the power and interest of many a good metaphor derives from how massively and conspicuously different its two subject matters are," David Hills writes, and this "comparison of two unlike things" is a general condition for all kinds of knowledge production.[44] We continually explore and expose discursive and formal connections in our fields, with all the potential and insights this entails. Nonetheless, comparison is a *theoretical* maneuver that requires caution. Data sets seldom call for the statistical, abstracting analyses performed on them, and often, similarities appear because a basis of comparison is created: because features are thought of and treated *as* similar. Moretti and others in the field of the digital humanities seem to be aware of this virtual level of comparison. However, examples of an opposite tendency also keep appearing, promoting digital methods to detect previously unknown patterns and trends in data sets,[45] asking us to "excavate" and "mine" data to discover novel insights.[46] It is this kind of modern empiricism that a project such as Stéphanie Solinas's may contribute to criticizing.

RECOGNIZING PATTERNS

Based on my analysis of the *Dominique Lambert* project and its framework of data processing, I identify (at least) two understandings of pattern recognition that play into and inform the absurdity addressed by Solinas. The first has to do with the obsession with formal similarity, known from pseudomorphosis: with the form and potential of abstraction, and with the hypothetical similarity that emerges when different or dissimilar subjects are brought together. In Charles S. Peirce's words, you could term this understanding of pattern recognition iconic, because it designates a kind of simulacrum in which an assumption (or, in Solinas's project, a demonstrative postulate) becomes the basis of the comparison, and where the individual Dominique Lamberts, each with their own particular preferences, are described and made comparable through their shared feature. The other understanding of pattern recognition has to do with an actual similarity between the participating Dominique Lamberts: with the compelling connection that emerges when the various individual features and preferences are merged. According to Peirce, this understanding of pattern recognition could be called indexical or indicative, because it designates a form of physical connection among the Dominique Lamberts. Here, we are closer to the actual data set, compared to the first understanding of pattern recognition, and the Dominique Lamberts have the opportunity to negotiate the similarities they may share. If the first understanding could be related to pseudomorphosis, the second understanding could, to use a term borrowed from psychiatry, be related to apophenia. Apophenia refers to an "abnormal connectedness between seemingly unrelated meanings,"[47] and in this context this may be thought of as the compelling and nagging feeling that in a strange way, the various Dominique Lamberts almost resemble each other, even though there is no reason to believe so. Although the first of the two forms of pattern recognition may be connected to a kind of a priori categorization that depends on a possible connection, the second one depends on an a posteriori comparison based on the subjects that are actually available.

My tentative formalization in the paragraphs above builds on a more extensive list of trichotomies developed by Peirce in his written correspondence with the English linguist Lady Welby during the winter of 1908.[48] The basis of these ten trichotomies is a fundamental tripartite of the sign based on Peirce's idea of three essentially different universes or three Modes of Presence,[49] including the Immediate, the Direct, and the Familiar, or firstness, secondness, and thirdness, as Peirce also calls them.[50] Above, I refer to firstness and secondness accordingly: to iconic and indexical ways of thinking about modern data science. Whereas firstness describes ideas or positive possibilities, secondness refers to things and their qualities. Thirdness, which I now add to the two forms of pattern recognition, involves mediation: it is the category of thought, language, representation, and it relates to the very process of semiosis, of making social communication possible.

While it would take too long to go into the richly detailed combinatorial modeling of Peirce's theory of signs, it is interesting to relate the two forms of pattern recognition in Solinas's project—and in data processing more generally—with Peirce's concepts of firstness and secondness, because it draws them into a more general methodological discussion. Based on this schematic listing, the first form of pattern recognition (the one pertaining to pseudomorphosis) may be related to firstness: to the possible or hypothetical, and the second (the one pertaining to apophenia) may be related to secondness: to the actual and concrete, or to an inductive way of thinking.[51] Here, the idea of firstness as a fundamentally possible or potential mode of presence is equal to the virtuality I described above. On this level, the basis of comparison is ambiguous and uncertain, and as a result, the comparison itself be-comes hypothetical: it contributes to understanding something that is not yet developed, something absent or abstract. On the other hand, secondness designates a mode of presence that belongs to a striking similarity between actual, available data points. Whereas firstness is related to abduction—to an experimental explanation of a hypothet-ical connection—secondness is related to induction—to a categorial explanation of actual conditions. These two understandings of pattern recognition may be related to each other and described as two central tendencies in modern data processing, as I do in the paragraphs above—one that sets up predefined categories and wrangles the data set into a useful form, and one that claims to excavate or mine the data set to discover and display its immanent and previously unknown patterns. Both tendencies build on and operationalize a certain way of thinking: they create connections and cohesion; "thirdness," in other words.

"Thirdness pours in upon us through every avenue of sense,"[52] Peirce poetically states in one of his Harvard lectures, and the same idea seems to haunt Solinas's project. Hypothesizing on the one hand and mapping on the other, both forms of pattern recognition build on general paradigms and ways of thinking that intervene in and compli-cate the matter in question. The project's Dominique Lamberts do not just come together in the data set, they are *set* there in the first place, brought into contact with one another based on particular interests and anticipations. Solinas makes them comparable, along with her actual comparison of them. Literary theorist Jeffrey R. DiLeo describes the relationship between firstness, secondness, and thirdness in the follow-ing way:

> [U]nlike secondness and firstness which are merely experienced and non-cognitive and incapable of being known, thirdness is cognition, viz., it is the mode of being of that which is such as it is in bringing firstness and secondness into relation with each other. Thirdness is the category of connection and mediation between firstness and secondness, but is not reducible to either of them. Thirdness may be characterized by the term's meaning, representation, mediation, and thought, although is best consid-

ered as both generality or universality and law: notions which are necessary for thirdness.[53]

Firstness and secondness are noncognitive and rely on experience rather than thought, according to DiLeo. In contrast, thirdness describes the connections and mediation between them, that is, the universality and law that enable their interrelation. In data science—according to Kitchin, for example—attention is brought to firstness and secondness: to patterns and regularities in data sets, and to how they may be detected and classified systematically.[54] However, that does not mean that the methods used to discover and describe these patterns are not characterized by a generality, that they themselves do not mediate when they connect and cohere.[55] Pushed to its logical conclusion, perhaps the very generality that enables pattern recognition in the first place is the overlooked yet is one of the most vital matters of concern when dealing with modern data science. You could ask, along with Kitchin and Peirce, whether data-driven science is as data-driven as it pretends to be or whether thirdness also inevitably "pours in" here. Is data-driven research not also an object for imagination and great expectations, dreams of the all-encompassing capabilities of innovative systems for the oversight of today's information deluge?[56]

Via Peirce's conception of firstness, secondness, and thirdness, I believe that you can more closely approach the way of thinking that pervades both abductive and inductive data processing and that contributes to bringing firstness and secondness together in the first place. This mediating aspect is often overlooked in the practical labor of data processing,[57] perhaps because data engineering and data science are fast-moving, competitive fields that are expected to generate profits, rather than scientific "truths" or explanations.[58] However, the mediation level is a decisive part of understanding how conventions and conceptions more generally contribute to and form the process of knowledge production: in other words, how thirdness mediates between firstness and secondness and how this mediation is part of the theoretical object of the data-driven investigation. In extension of this, it is relevant to further dive into the idea of thirdness as a middle or an axis of mediation. Whereas firstness and secondness take the form of a dyad, a beginning and an end, cause and effect,[59] thirdness marks the continuity between the two and displays the course from *A* to *B*, while introducing the middle as a presumed position: thirdness is not just a simple channel, a route or wire through which a signal flows, but an aesthetic focal point that generates meaning as such.

When attending to the aesthetic level of Solinas's project, one may ask, "What is it that we see *with* or *through* when we conduct data-driven research, and which formal categories are established, as we undertake experiments that are supposed to be scientifically rigorous?" Along with Lorraine Daston, one could also ask, "What scientific 'before' affects the knowledge production of the scientific 'now'?" "When and how does empiricism become *collective*?"[60] Dominique Lambert is described and constructed through numerous processes of comparison and translation, and so we witness mediation techniques as varied as psychological surveys, sketches, simulations, and classical portraiture. The passport photographs of the "real" Dominique Lamberts close the chain of representation, we are told,[61] just as the intermediate stage of the process is emphasized: experts from the Comité Consultatif pour la Description des Dominique Lambert are brought in, and professional equipment is borrowed from the police's Criminal Identification Bureau.

This aesthetic construction of Dominique Lambert as a figure emerges as both a formal similarity (as seen through the lens of pseudomorphosis) and an actual contiguity (as seen through the lens of apophenia) between the participating Dominique Lamberts as both abstraction and connection. Dominique Lambert is produced theoretically in the project, is *thought* into existence, you might say.

DOMINIQUE LAMBERT AS A PATTERN

Throughout the hundreds of pages of the *Dominique Lambert* book, through five translation sequences consisting of comparative questions, and written, drawn, computer-generated, and photographed portraits, Solinas examines and dissects the concept she has developed with this project: Dominique Lambert as both a category and as a multitude of receding chins, turned-up noses, and greying temples; poorly executed computer simulations of jawlines, necks, and earrings. The Dominique

Lambert project is primarily an aesthetic investigation into the nature of comparison. It is an exploration of what comparison might look like, if offered a visual language that mimicked (or parodied) its systematic structure, and of what happens when an absurd research concept is followed through systematically.

However, the Dominique Lambert project may also be seen as more than just the aesthetic concretization of a complex question. As it captivatingly and accurately visualizes its problem, making it accessible for examination and discussion, it also produces it through its conceptual and methodologically rigorous structure. The project is not only about comparison; it also compares itself. It operates and intervenes in the field around which it thematically revolves, and this parodic form contributes to making it dynamic. Comparison not only requires bringing together two or more things, it presupposes comparability in the first place.

It appears that Solinas is acutely aware of this, as she has carefully made her project comparable to what she investigates, and this homogenizing act of making comparable is crucial to the rationale behind it. Through aesthetic examinations of the nature of knowledge production, including the aesthetic logic of Chinese portraits, the transformative function of translation (not just from one word to another, but also from one medium to another), and the ethos and weight of the expert evaluation involved, Solinas not only illuminates how comparison appears as a mode, but also how it operates. The *Dominique Lambert* project may be seen as an opportunity to discuss some of the methodological questions related to ideas of comparison. Comparison as a mode of operation pertains to various types of humanistic knowledge production, particularly comparatism as a methodological strategy in art and literary analysis. One of the central questions raised by the *Dominique Lambert* project is the one of similarity, or more precisely, how similar something needs to be to something else for one to reasonably assume a form of kinship. Deleuze's concept of virtuality causes a series of connections or potential points of linkage to emerge, which could not otherwise be understood. At the same time, there are also possible concerns associated with comparison if the individuality of the examined subjects erodes. Comparisons are artificial constructs, and if they are based solely on assumed sameness, the comparative analysis risks obscuring its own theoretical status.

In this chapter, I have examined and described the two overarching types of pattern-recognition logic at work in Stéphanie Solinas's artistic project, *Dominique Lambert*, and more generally in contemporary data processing. These types of logic include an abstracting a priori form of pattern recognition (pseudomorphosis) and an inductive a posteriori form of pattern recognition (apophenia). I have used Peirce's ten trichotomies and the three fundamental Modes of Presence of his phenomenological system to trace the recurring figure around which both pattern-recognition forms revolve, namely, Dominique Lambert as a product or questioning strategy. I noted the inherent thirdness that emerges in the project's numerous translations, comparisons, and mediation, as I argue that information and knowledge do not simply emerge from either the analytical concepts or the data set, but instead arise in the interaction of both a framework and a set of conventions.

In this chapter, I have argued that a new object of knowledge emerges from new data processing methods and automated systems, including machine learning and other types of artificial intelligence. New classifications become possible, and new virtual connections may be created. Drawing on Rob Kitchin's critical diagnosis of today's empirically oriented scientific environment, and aided by Solinas's eccentric and incisive art project, I highlight how new types of data processing not only yield comparisons, connections, and patterns, but also generate new ideas, mindsets, and a new digital object to which they refer. In accordance with the main concept of this book, the digital object and the concepts of comparison and comparability are crucial. Although comparison as a mode and topic is far from new—just think of the comparison between Punic sarcophagi from the third century BCE and tombs from the High Gothic period, 1,500 years later—and although the nature of comparison is well described in both literary and art theory, something crucial happens when this ancient cultural technique is electrified, so to speak. Comparisons of two (or more) subjects suddenly take on an entirely new scope, as the qualitative is supplemented with the quantitative, the near with the distant,[62] and two becomes two hundred or—in the completely detached, data-driven version—two hundred thousand or two million. Quantity and scale are debated topics at this time,[63] and we find ourselves in the moment of the archive, a point in time where at one moment we imagine marvelous, Borgesian libraries with infinite shelf space, and at the next, we have nightmarish visions of irretrievably losing all the data and information we have collected and stored.[64]

To handle, process, and make sense of large amounts of data, they must first be *made* comparable. They must be reduced to a set of pre-filtered parameters by which they may be compared, abstracted, and made "alike," or they must be algorithmically examined with an eye for inherent numerical patterns; points closer to or farther from each other than initially believed. Comparison has always yielded objects of knowledge—peculiar, theoretical entities that establish connections and relationships where there were none, one might argue. However, when examined as a specific digital mode, comparison is characterized by a different radicality than previously. If the comparison of two subjects blurs the contextual difference between them in favor of an emerging similarity, the digital comparison of two hundred thousand or two million subjects does so even more profoundly. Here, it is not just a matter of one-dimensional similarities and differences between two subjects, but of formal, numerical patterns in many dimensions that cut across individual measurements and contexts. It is simply easier to preserve the contours of a contextual thirdness if two subjects are compared, instead of two hundred thousand, just as it is easier to grasp the consequences of the folding of differences and untranslatability if the data set has not first undergone an often irreversible reduction and abstraction for the sake of comparability.

Conclusion

A DATA-SATURATED WORLD

••• In just two decades, our world has undergone a remarkable transformation, one that is underpinned by the pervasive influence of data processing and analysis. From the moment we wake up to when we lay our heads on our pillows, our daily lives are immersed in a digital landscape where data-driven insights shape our experiences. Online shopping presents tailored product recommendations, and social media orchestrate feeds that brim with images and content, fine-tuned to our interests. Streaming platforms anticipate our entertainment desires, and apps forecast the weather with unprecedented accuracy. The guidelines for our health and fitness are mapped out through personalized analytics, and financial decisions are guided by data-rich insights. We navigate traffic-laden streets with real-time updates and find love through algorithms. From graphs and charts that enhance news articles, to the accurate analysis that powers video games, from healthcare portals that display our vital statistics, to market research graphs that shape consumer choices, the impact of data analysis is ceaseless and profound. Data processing guides us and influences what may be seen, imagined, and discussed—shows us patterns, connections, and correlations that were previously invisible. As we immerse ourselves in this data-saturated state, our perceptions of reality are fundamentally *molded* by the algorithms that curate our news feeds, the personalized recommendations that shape our preferences, and the visualizations that elucidate complex concepts.

In this book, I have explored the background of the current data-driven paradigm and discussed how elusively, yet crucially, statistical and computational ideas and concepts form the way data have been and are experienced and made sense of historically and in the present. To do so, I have introduced the concept of the digital object to describe, connect, and make sense of some of the many disparate ways which data processing works in our digital culture today. From the very first pages of this book, I have contended that the digital object may be seen as a necessary tool for theoretically delving into and analytically grasping the heterogeneous and complex data-driven environment we encounter. I have argued that the digital object is not a physical entity but should be understood as being akin to other theoretical constructs, such as abstract-formal language systems[1] or political economy.[2] It cannot be measured, weighed, collected, scrutinized, or observed. It does not exist "out there." Instead, it may become the subject of analysis when it is concretized with the aim of revealing something that was otherwise opaque, assembling something that was previously scattered, or comprehending something that otherwise eluded understanding. I have explored various manifestations or realizations of this concept. Throughout this book's chapters, I have focused on three distinct modes of this way of thinking, or three variants, so to speak: excluding (outliers), combining (aggregates), and comparing (patterns). Each chapter has examined the nuances of these modes and illustrated the ways in which the digital object takes form in each of the three contexts. These modes of operation are not abstract groupings intended to categorize the digital realm into three main categories. Instead, they represent three actual contexts where data processing is utilized to address politics, society, and aesthetics. I do not claim that the preceding pages have presented an exhaustive, stable, or finished depiction of our data-driven and digital culture in its current form. Although the concept operates on a theoretical level, formalizing and abstracting the specific phenomena it seeks to describe, the aim of this book has never been to systematically map out or categorize all the various forms of digital artifacts. Nevertheless, through my analyses, I have attempted to approach what could be seen as a conceptual and theoretical synthesis of what the digital may entail.

THE DIGITAL OBJECT

In the introductory chapter, I described how the concept of the digital, which stems from the etymological root, *digitus*, has been and continues to be perceived as a fundamental logic of counting. Historically, the term *digital* has been associated with questions that concern data and information as representations of signals, whether magnetic or electrical. Additionally, it has often been contrasted with the term *analog* to emphasize this representational connotation. Next, I argued that the act of counting or enumerating the things around us has generally been linked to a derivative mode of operation where representations of objects, rather than their actual presentations, have taken precedence. In other words, meaning is abstracted and drawn away from things so their symbolic value may be retained. Building on this premise, I suggest that such a derivative or mathematical mindset should be complemented with a situating, contextualizing, and process-oriented approach. This approach considers the specific space, the concrete aesthetic practice, and the unique culture in which the digital operational method unfolds. Thus, when the digital is described in this book, it is done with the intention of understanding the nature of digital operations and computations, and how they concretely impact, renegotiate, and create social, cultural, and aesthetic concepts. I suggest that the digital should be understood as dual-natured: it is specifically directed at the particular contexts in which it is realized and simultaneously denotes a synthesized understanding of the diverse array of ways in which this realization manifests.

The distinctive digital logic of counting serves as the focal point for the three chapters, where it interweaves the analyses like an ever-present figure throughout the practices it informs and is informed by. In the first chapter, the mode of counting encompasses selection and classification as it emerges as a fundamental form of exclusion or omission. Concrete strategies that homogenize data sets are employed to moderate and sort the outputs of investigations. In the second chapter, the process of counting emerges in the figure of the informational aggregate. That chapter examines the counterintuitive idea that one may indeed gain information by discarding information, namely, the individuality of observations. Here, for instance, details about individuals included in a large demographic study are ignored in favor of more fundamental or deep properties. In the third chapter, the counting logic resonates as a mode of comparison used to unveil patterns and connections. Hence, the counting logic operates not merely as a simple exercise of "counting on fingers," but prompts a range of far more intricate operations, with significant aesthetic, cultural, social, and political consequences. As I describe how the digital object engages with entities in grammatical, phenomenological, and dispositive terms, these are the consequences to which I refer throughout this book.

In grammatical terms, the digital object behaves similarly to deictic terms—those such as *here, now,* or *this*—which rely on context.[3] These deictic terms serve as examples of how context and cognitive orientation affect meaning. Similarly, the digital object has a spatiotemporal context and indicates the practices to which it pertains. In the introductory chapter, I present listing as an operation, and the list as a thing, to illustrate how a theoretical, digital object depends on a specific context. Lists never simply convey information, they also demarcate boundaries between distinct entities and inscribe a specific sorting logic.[4] They transform people, words, or things into dynamic entities that may be processed, stored, or transmitted; they determine which words are significant and which are lexically redundant,[5] and they inscribe objects into the symbolic order, rendering them amenable to manipulation, revision, or deletion.[6] However, listing also implies a theoretical digital object that takes the list as its subject, so to speak. The list enables alphanumeric inscription into the symbolic order; it shifts, transforms, and removes individuals and things, rendering them compatible with the structures and standards of the investigation and the enumerative means employed. This process highlights how the list, as the object of the listing mode, is crucial for it to achieve cognitive, symbolic, or political significance. As a necessary but ephemeral anchor for digital thought, the list, as a logical construct rather than a material entity, becomes part of knowledge development. This is closely intertwined with how the digital object may, phenomenologically, be said to adopt or assume a provisional form, as its abstract logic is translated into, or realized, through events, actions, and experiences that frame the context of the knowledge development. As I argue in the introductory chapter, this is not solely a matter of a logical, mathematically representative relationship, but of a more intricate, contextually defined and mediated connection between the world and the ways it presents itself to us—for instance, via various technological representations. When we create new theoretical objects, we also introduce novel ways of perceiving the world—new lenses through which objects and relationships may be examined, new distributions of the visible and the invisible. The same applies to the digital object as a theoretical entity. It does not exist in the usual sense of the word, yet it shapes and transforms the numerous phenomena it touches, offering itself as a lens through which they may be observed, or as a path along which they may be conceived and conceptualized. The digital object meshes with and predetermines what we can see and comprehend, and offers new ways of systematizing and regulating. As a theoretical mediating apparatus, it interacts with a specific context, and articulates and establishes new differences in, resemblances between, and distributions of the visible and the invisible.

The three ways in which the digital object takes subjects as its things, or becomes concrete in a specific context, are all linked to its effects, viewed from an aesthetic, cultural, social, or political perspective. I suggest that the digital object cannot be reduced to the materiality of wires, cables, and flickering pixels, nor can it be comprehended merely as an abstract computational process detached from use and context. Instead, it reflects a distinct way of thinking or perhaps more precisely, what we think *with* when we execute digital calculations and visualizations. I have carried out artistic and philosophical investigations to demonstrate that statistical and computational concepts do not appear only aesthetically and culturally, or take shape as unintentional or superfluous side effects of a more general way of thinking. Instead, they are entirely reliant on aesthetics and mediation to acquire significance in knowledge development. I argue that data-driven knowledge development is fundamentally situated and contextual, whether it occurs in large-scale demographic studies, data mapping, or comparative analyses. Drawing on Peirce's concept of the mediating thirdness, I integrate this point into analyses of today's data processing methods. I argue that digital phenomena cannot be reduced to machine components (firstness) or formal, algorithmic relationality (secondness), but are always encountered via a mediator or along a continuous axis between the two. Thirdness is not merely a conduit through which a signal or specific content can flow; instead, it is the focal point of the production of meaning.

In extension of the foregoing argument, I note a less abstract interpretation of the term *digital*, which is rooted in the etymological meaning of *digitus*, which we encounter in the form of *deuk* (including the Sanskrit *dic*, meaning "to show or point out," the Greek *deiknynai*, meaning "to show or prove," the Greek *dike*, meaning "common use," the Latin *dicere*, meaning "to speak, tell, or say," and *zeigon* and *zeigen*, from Old High German and German, respectively, both meaning "to show"). Although this sense is often absent from the understanding of the term *digital*—which is typically approached as an information theory concept—it is still etymologically traceable. When I first wrote about such a "deictic" orientation, and highlighted it as a kind of counteraction to an isolated mathematical-technical interpretation, it was to direct attention to the aesthetics of the digital, to the concrete conditions that constitute the far-from-isolated data representations of the surrounding world. My exploration of the digital "counting on fingers" does not point away from or outward to an abstract process or algorithmic mode of functioning, but specifically to the setting in which it all happens. The digital object always materializes within a specific setting, in a concrete aesthetic practice in digital culture, just as its use always presupposes an anchor and a necessary place from which analysis may begin.

AESTHETIC AND CULTURAL PERSPECTIVES

The aesthetic and cultural perspectives associated with each of the three computational and statistical concepts, represented in their respective chapters, interact with the overarching way of thinking I describe through the digital object. Thus, each of the three chapters contributes in its own way to the discussion of aesthetic and cultural knowledge development. However, each chapter also presents its own insight, and next, I will briefly outline these to ultimately connect them to the broader research perspectives they indicate.

The first chapter of this book focuses on individuals who, for one reason or another, do not fit into, or cannot be adapted to, the prefiltered categories of data processing. This chapter delves into the aesthetics of noise, deviations, or what is generally rejected—what lies beyond the immediately usable, interesting, or acceptable. I draw inspiration from artist Rossella Biscotti's many textile works, and the systematic demographic surveys from which they derive, to explore ideas of norms and normality, and negotiations of what falls outside of them. I highlight how the "normal" (in the statistical, economic, cultural, and social sense of the word) cannot exist without the anomalous and reference the totality paradigm upon which large-scale studies are built.[7] Classification must be comprehensive to make sense, and the set of categories established through systematic mapping must include all the units encompassed by a study. If something does not fit in, it is still categorized under labels such as "other," "others," or "miscellaneous." Thus, knowledge development relies on a degree of virtuality, where one or more stand-ins or placeholders that encompasses the uninteresting, disruptive, or noisy aspects of a study are necessary. This point is interesting not just from a statistical perspective; it also sheds light on broader understandings of usability and noise, and gives rise to discussions that are inherently aesthetic and cultural: Who is considered part of the "normal" group or category, and how is this visualized? What is the intent of the study, and how is its purpose achieved? Knowledge development is situated and concretely mediated, and always has a political, economic, and social context.[8] It interweaves strategies and objectives, as well as power dynamics and discourses.[9]

The second chapter of this book builds on Adam Broomberg and Oliver Chanarin's conceptual explorations of their selection and combination of CCTV photographic material from systematic, automated surveillance to examine various concepts of data transformation. This chapter explores how the human face is first rendered manipulable through symbolic encoding and how the nonquantitative is made quantifiable through numerical modes of operation, such as aggregation. This chapter examines what it means to transform and quantify, to derive meaning from extensive data sets, by referencing sources that include the historical mapping of the rod as a unit of measurement, Francis Galton's composite photography, and August Sander's photographic investigations of the German population during the Weimar years. The underlying aesthetic endeavor scrutinizes what is wanted and what is sought when one combines and selects. A key argument is that the final aggregates—the coolly detached and contextually stripped-down facial maps—reveal the same intentionality that they embody. This kind of systematic data mapping seeks something beyond individual measurements, something that emerges as the data set is organized. This intentionality manifests as imperfections and glitch-like cracks in the images' expressions. These aggregates were never intended as portraits of individuals; instead, they function as biometrically and numerically suitable representations of citizens associated with a certain social (e.g., criminal) category. They reveal the thought process that gave rise to them, and their visual expressions indicate their purpose—to statistically determine identity in relation to a database with numerous other entries to which they may be linked. Similarly to Galton's studies of the average human being among groups of siblings, convicted criminals, or individuals with the same illness, the surveillance algorithms in Broomberg and Chanarin's artistic project seek the statistically significant facial features when biometrically distinguishing individuals from one another. This enables systematic processing by authorities and private entities alike. The artworks highlight the quest for the statistically average traits of importance, just as Galton sought to do with his composites. The intention, the approach, and the implications converge, revealing how the visual language reflects the mindset and purpose, shaping the identity and understanding of data subjects within the broader data processing landscape.

The third chapter also encompasses its own aesthetic and cultural perspectives, and delves into data processing and digital knowledge development, and the age-old cultural technique of comparison. Anchored in the artist Stéphanie Solinas's extensive mapping project, *Dominique*

Lambert, this chapter examines how data-based patterns are created through a series of translation, comparison, and mediation processes, unveiling the underlying digital logic from which they stem. Here, the mediated nature of knowledge development is demonstrated through the systematic production of similarity—or the basis for comparability, as I call it in this chapter—and I delve into the specific processes that constitute this concept of similarity. Even though comparison is not a new concept, I argue that something crucial occurs when it is electrified, so to speak. I note the perspectives of fields such as the digital humanities to highlight how comparing two subjects assumes an entirely new scope when the qualitative is supplemented with the quantitative, turning two into two hundred. I argue that the humanities' fundamental concept of comparison must be reimagined in a digital and data-driven context. Just like many other concepts, it is transformed and adapted because of the digital and algorithmic counting logic that pervades today's digital culture.

The chapters of this book offer relatively diverse perspectives on how the digital object could be accessed and understood through three concrete realizations: exclusion (outliers), combination (aggregates), and comparisons (patterns). Instead of attempting to axiomatically establish a set of conditions for, or requirements of, the digital, I adopt a more pragmatic approach. My aim is not to exhaustively cover all the digital modes I propose, nor do I attempt to narrow down or unequivocally define what the digital is. Digitality manifests across various platforms and in a multitude of contexts. This book's three chapters could identify many other phenomena that fall under the overarching themes of exclusion, combination, and comparison, just as there may be other overarching themes that could be explored through the theoretical framework I propose. The way I conceptualize data processing in this book is just one of many possible ways, and the three digital modes I have chosen to investigate represent merely a fragment of a much larger field that is rapidly evolving these days.

A unifying thread runs through the three case studies I present, a shared logic that explains why I have chosen to focus on them specifically. What binds these chapters together is their common foundation in aesthetics, an examination of the process of how knowledge is developed and presented. My aim in these chapters is to elucidate and challenge prevailing beliefs about digital knowledge production and data processing, which is underpinned by a fundamental idea: the very act of knowledge becoming visible is of paramount importance. Questions arise, such as, What may be understood through statistics and data processing? What characterizes a computer scientist's knowledge, versus that of a statistician—and what tools do they employ to reach their understandings? Next, in my concluding reflections, I will endeavor to provide a more comprehensive exploration of this consistent theme, which underpins the entirety of this book.

AESTHETIC INVESTIGATIONS

This book's three chapters represent three specific sites in digital culture where knowledge production emerges through a series of sensory conditions. Simultaneously, the perspective from which I examine these contexts in these chapters may also be described as aesthetic, as I explore the resources that a sensorially grounded way of thinking may offer in relation to comprehending and complicating my subject—not only in terms of my close readings of specific artworks, but also for science and knowledge development more broadly.[10] However, what does it mean to inquire into these highly technical, statistical, and information-theory operational methods from the standpoint of aesthetics? How does this particular approach contribute to the understanding of current digital culture?

Several times throughout this book, I describe how the artistic practices I included allowed me to operate analytically on both a concrete, specific level, and an abstract, formal level. For instance, when studying larger demographic mapping, I first investigated how a particular artistic project operated critically and aesthetically in its thematic setting, and then more deeply examined in the specific operationality and thinking produced by the punched card technology and its computational successors. I am also concerned with the more general or overarching mode through which knowledge comes-into-being, when one thinks with, and develops, theoretical concepts along this artistic line of thought.[11] In the case of demographic mapping, in the second chapter, I am interested not only in two- and three-dimensional facial recognition software, but also in the systematic way of thinking it represents, and its historical and theoretical precursors. I am interested in the concrete setting—in the encounter with the strange and remote computational "face masks"—but I am also interested in the ideas and ambitions they are a result of. Together, I believe that these two levels provide a solid foundation for examining a range of phenomena that may initially appear mathematical, technical, and statistical, but nonetheless offer themselves to more fundamental epistemological discussions when approached from a critical aesthetic perspective. Thus, the analytical and theoretical foundation of this book rests on the premise that examining computational modes through an artistic and aesthetic lens is not only beneficial but also necessary. It asserts that the development of knowledge, whether theoretical or statistical, is inherently intertwined with aesthetics; that iconic and symbolic elements are intrinsically connected, borrowing from Peirce's terminology once again.

You may argue that my conception of the digital object as a theoretical object indicates a fundamental diagrammatic level in the computational thinking I analyze. Peirce's understanding of the diagram fundamentally rejects dualistic forms of thinking (intellect versus observation, cognition versus iconicity, subject versus object) and the pervasive idea that mediation shapes and reveals knowledge as it emerges through a dynamic interplay between conceptual and instrumental tools, and human perception and cognition. When I suggest that the digital object may be explored through Peirce's idea of the diagrammatic, it is specifically to emphasize this dynamic nature.

I insist on using the term *object*, because I believe that the digital modes of operation studied in this book assume an object-like character as we use them to think through or along. However, I am also aware that this choice of words may inadvertently steer thoughts toward more dualistically grounded thinking: an object often relates to the more dominant subject, which, unlike the object, has the capacity to act and enact. Nevertheless, I do not subscribe to such a "modern"[12] understanding of the object in this book. Instead, I believe that as a conceptual tool, the digital object shapes and cocreates the information and knowledge we encounter in connection with new forms of data processing—and vice versa. I understand the digital object as a theoretical intervention in a complex field, which does not necessarily expose and clarify all patterns and connections, but instead offers an entry point, a lens, or a conception through which previously hidden patterns emerge. Not only is this process shaped by its context, but it, in turn, shapes what is being investigated. The subject (in the modern sense) is not the only agent at work.

Theoretical concepts and ideas are far from passive entities that merely wait on the whims of human agents. Such a theoretical standpoint is also embraced by A. S. Aurora Hoel in her article on Peirce's diagrammatic thinking titled "Lines of Sight: Peirce on Diagrammatic Abstraction." In this piece, which draws inspiration from Peirce's work, she focuses specifically on the transformative potential of diagrams and discusses a new dynamic and differential approach to mediated knowledge development:

> [D]ynamic, since it conceives meaning and knowledge in terms of open-ended processes of articulation and identity in terms of progressive processes of coming-into-form; and *differential*, since it emphasizes the transformative powers of mediating apparatuses and the way that they, in their very act of articulating, introduce new divisions, new distributions of the visible and invisible, provoking phenomena to grow, so to speak, beyond themselves.[13]

She emphasizes that concepts also mediate (and are mediated by) what they describe, a process that is both unlimited and dynamic, and inherently transformative. Put differently, the lines along which we see matter, as each "media matrix,"[14] as she terms it, sets in motion its own sphere of visibility. Lines distinguish, divide, and distribute—they make visible and invisible, highlight and diminish. Therefore, these lines or diagrammatic matrices tend to assume object-like characteristics, giving rise to new lines of sight. She writes, citing Peirce: "[O]ne and the same construction may be, when regarded in two different ways, two altogether different diagrams; and that to which it testifies in the one capacity, it must not be considered as testifying to in the other capacity."[15] In other words, the forms and lines along which thinking emerges are crucial to shaping how diagrammatic abstraction may be fashioned.

The foregoing two levels of concreteness complement each other: the forms the digital object takes matter—which lines are drawn and what shapes emerge matter. Yet, at the same time, this particular realization must not be confused with the digital object as such. This realization is just one of the many forms it could take. In the second chapter, I approach the sculptural data mapping aesthetically and draw attention to the correspondences between the two levels: the concrete embodiment of the incomplete composition of shifting surfaces, and the algorithmic operationality that extracts and discards information, evaluating the gazes and gestures of the subjects portrayed. Making visible and invisible, amplifying and muting, Broomberg and Chanarin's data mapping works on both a concrete and a theoretical level.

Although the digital object lacks a tangible form, it makes it possible to describe the digital logic and computational modes of operation with and through which we think when we undertake theoretical and practical work today. It has the ability to transform an experience of the world into an unambiguous digital representation and to share these digital data with various communities of knowledge, or apply them to politics or economics. Stated in a slightly different way, you could say that lines of sight, in the sense Hoel proposes, create meaning at a far more profound level than just the visual, much like knowledge development—whether theoretical or data-driven—must be thought of as a fundamental aesthetic concern; the symbolic, as constantly intertwined and interconnected with the iconic.

NOTES

Introduction

1. Lorraine Daston, "Why Statistics Tend Not Only to Describe the World But to Change It," *London Review of Books* 22, no. 8 (2000), 35–36.

2. The law of large numbers, first propounded by Italian mathematician Jacob Bernoulli in 1713, and further described by French mathematician and physicist Siméon Denis Poisson in 1837, suggests that as a sample size grows, its mean moves closer to the average of the whole population. This idea has had a profound influence on statistics, particularly in relation to its applications to sociology and political economy in the nineteenth century, a discussion I go into in chapter 2.

3. The Gaussian distribution, also known as "the normal distribution," is a widely used model for the distribution of continuous variables, e.g., Christopher M. Bishop, *Pattern Recognition and Machine Learning*, Information Science and Statistics (New York: Springer, 2006), 78–79. Although some credit the mathematician de Moivre with formulating the normal distribution, it is most often associated with the German mathematician and scientist Carl Friedrich Gauss, who introduced this important statistical concept in his 1823 monograph, *Theoria combinationis observationum erroribus minimis obnoxiae* ("Theory of the Combination of Observations Least Subject to Error"), where, among other things, the normal distribution is formulated. See Carl Friedrich Gauss, *Theory of the Combination of Observations Least Subject to Errors,* trans. G. W. Stewart, Classics in Applied Mathematics (Philadelphia: SIAM, 1995).

4. In his 2014 article, Rob Kitchin examines what he identifies as the new "empiricist epistemologies" of modern science, which "declare 'the end of theory,' the

creation of *data-driven* rather than *knowledge-driven* science, and the development of digital humanities and computational social sciences that propose radically different ways to make sense of culture, history, economy and society" (my emphasis). See Rob Kitchin, "Big Data, New Epistemologies and Paradigm Shifts," *Big Data & Society* 1, no. 1 (2014): 1. He argues that the emergence of Big Data and new data analytics are disruptive innovations, which reconfigure how research is conducted, and he calls for broader critical reflection on the part of the academic disciplines working with data processing concerning the epistemological implications of the developing data revolution. This book is intended to contribute to such a critical endeavor.

5. Liv Hausken, *Thinking Media Aesthetics: Media Studies, Film Studies and the Arts* (Frankfurt am Main: Peter Lang International Academic Publishing Group, 2013), 31.

6. As Lotte Philipsen and Rikke Schmidt Kjærgaard state, "Representation, scientific data representation, too, is inevitably a matter of aesthetics since all representations are created and shaped under human influence. Even a graph on a monitor has at least shape, color, position, and size, and however advanced devices we may use in the process, these representational features are culturally and technically constructed by humans"; see Lotte Philipsen and Rikke Schmidt Kjærgaard, *The Aesthetics of Scientific Data Representation: More Than Pretty Pictures* (New York: Routledge, 2018), xii.

7. A. S. Aurora Hoel [formerly Aud Sissel Hoel], "Lines of Sight: Peirce on Diagrammatic Abstraction," in *Das Bildnerische Denken: Charles S. Peirce*, ed. Franz Engel, Moritz Queisner and Tullio Viola (Berlin/Boston: De Gruyter, 2012).

8. This book draws on an understanding of aesthetics that originates in the concept introduced in 1735, when German philosopher Alexander Gottlieb Baumgarten defined it in his master's thesis, *Meditationes philosophicae de nonnullis ad poema pertinentibus* [Philosophische Betrachtungen über einige Bedingungen des Gedichtes], to mean *epistêmê aisthetikê*, or the science of what is sensed and imagined; see Alexander Gottlieb Baumgarten, *Reflections on Poetry: Alexander Gottlieb Baumgarten's Meditationes philosophicae de nonnullis ad poema pertinentibus*, ed. Karl Aschenbrenner and William B. Holther (Berkeley: University of California Press, 1954), 86–87.

9. Although the idea of an aesthetic "attitude" has been hotly debated, particularly as it is understood in aesthetic attitude theories (e.g., George Dickie, "The Myth of the Aesthetic Attitude," *American Philosophical Quarterly (Oxford)* 1, no. 1 [1964]; for a response to this critique, see G. Kemp, "The Aesthetic Attitude," *British Journal of Aesthetics* 39, no. 4 [1999]). It is relevant to this topic to discuss how an observer addresses and experiences aesthetic phenomena. Therefore, the attitude adopted in this book is one of searching for aesthetic features in many kinds of objects, and not just art works or traditional aesthetic objects (or even *beautiful* objects). By focusing on aesthetic features of a scientific diagram or census-taking

model, for example, I seek to understand how knowledge is produced along the lines of particular instantiations of abstract thinking. For an in-depth discussion of the term *attitude* and its problems as philosophical term, see Alexandra King, "The Aesthetic Attitude," in *The Internet Encyclopedia of Philosophy*, https://iep .utm.edu/aesthetic-attitude/.

10. Jørn Erslev Andersen, *Sansning og erkendelse: Æstetikhistoriske grundtekster fra Baumgarten til Kant* (Aarhus: Aarhus Universitetsforlag, 2012), 237.

11. Hoel, "Lines of Sight."

12. Hoel, 268.

13. To paraphrase Michel Foucault's definition of the apparatus (*dispositif*), as quoted by Giorgio Agamben, "[t]he apparatus is thus always inscribed into a play of power, but it is also always linked to certain limits of knowledge that arise from it and, to an equal degree, condition it. The apparatus is precisely this: a set of strategies of the relations of forces supporting, and supported by, certain types of knowledge." Giorgio Agamben, *What Is an Apparatus? And Other Essays* (Stanford, CA: Stanford University Press, 2009), 2.

14. Agamben, *What Is an Apparatus? And Other Essays*, 2.

15. Douglas R. Harper, "concept (n.)," in *Online Etymology Dictionary* (Tupelo, Mississippi, 2021), https://www.etymonline.com/word/concept.

16. See Karl Marx and Frederick Engels, *Collected Works of Karl Marx and Frederick Engels*, vol. 28 (London: Lawrence & Wishart Electric Books, 2010), 49.

17. See Ferdinand de Saussure, *Course in General Linguistics*, ed. Charles Bally, Albert Sechehaye, and Albert Reidlinger, trans. Wade Baskin (New York: Philosophical Library, 1959), 7–8.

18. For a great introduction to Wolfgang Ernst's materialist approach to media theory and history, see Wolfgang Ernst and Jussi Parikka, *Digital Memory and the Archive* (Minneapolis: University of Minnesota Press, 2012). For an example of Jussi Parikka's materialist approach, see Jussi Parikka, *A Geology of Media*, Electronic Mediations (Minneapolis: University of Minnesota Press, 2015).

19. See Yuk Hui, *On the Existence of Digital Objects*, Electronic Mediations (Minneapolis: University of Minnesota Press, 2016).

20. *Kulturtechniken* has been translated into English in many ways over the years: "cultural technologies," "cultural techniques," and "culture technics" (with and without a hyphen). As Geoffrey Winthrop-Young mentions in his translator's note in Bernhard Siegert's book, *Cultural Techniques: Grids, Filters, Doors, and Other Articulations of the Real*, the greatest dilemma in this translation is the word *Technik*, as its connotations range from "gadgets, artifacts, and infrastructures all the way to skills, routines, and procedures—it is thus wide enough to be translated as technology, technique, or technics." Bernhard Siegert, *Cultural Techniques: Grids, Filters, Doors, and Other Articulations of the Real*, trans. Geoffrey Winthrop-Young (New York: Fordham University Press, 2015), xv. Winthrop-Young opts for "techniques," and manages to encompass "drills, routines, skills,

habituations, and techniques as well as tools, gadgets, artifacts, and technologies," as possible associations, and, arguably, "cultural techniques" is the most appropriate term. It is the one I use in this book.

21. This particular inflection of the term references its etymological roots in culture, and derives from the Latin *cultura*, which in turn derives from *colere* (meaning "to tend, guard, cultivate, till").

22. Geoffrey Winthrop-Young, Ilinca Iurascu, and Jussi Parikka, "Cultural Techniques," *Theory, Culture & Society* 30, no. 6 (2013): 5.

23. The quote is from a translated and revised edition of the original 2003 article published as "Kultur, Technik, Kulturtechnik: Wider die Diskursivierung der Kultur," in Sybille Krämer and Horst Bredekamp, eds. *Bild, Schrift, Zahl* (Munich: Fink, 2003): 11–22. The English translation, "Culture, Technology, Cultural Techniques—Moving Beyond Text," is provided by Michael Wutz and appeared in *Theory, Culture & Society* 30, no. 6 (2013), quote on page 27.

24. Krämer and Bredekamp, "Culture, Technology, Cultural Techniques—Moving Beyond Text," 25.

25. Cornelia Vismann, "Cultural Techniques and Sovereignty," *Theory, Culture & Society* 30, no. 6 (2013): 83.

26. Liam Cole Young, "Cultural Techniques and Logistical Media: Tuning German and Anglo-American Media Studies," *M/C journal* 18, no. 2 (2015).

27. Bernhard Siegert, "Cultural Techniques: Or the End of the Intellectual Postwar Era in German Media Theory," *Theory, Culture & Society* 30, no. 6 (2013): 61–62.

28. Michel Serres, *The Parasite*, trans. Lawrence R. Schehr (Baltimore: Johns Hopkins University Press, 1982), 13.

29. Siegert, *Cultural Techniques*, 23.

30. See Siegert, 23.

31. See Young, "Cultural Techniques and Logistical Media." See Cornelia Vismann, *Files: Law and Media Technology*, trans. Geoffrey Winthrop-Young, Meridian, Crossing Aesthetics (Stanford, CA: Stanford University Press, 2008), 5–6.

32. See Daniel Rosenberg, "Stop, Words," *Representations* 127, no. 1 (2014).

33. See Young, "Cultural Techniques and Logistical Media." See Sybille Krämer, "The Cultural Techniques of Time Axis Manipulation: On Friedrich Kittler's Conception of Media," *Theory, Culture & Society* 23, no. 7/8 (2006).

34. Here, "politics" is closely related to the sensorial, that is, to the realm of aesthetics. Politics as an aesthetically mediated phenomenon may be understood in relation to the definition proposed by Jacques Rancière: "This aesthetics should not be understood as the perverse commandeering of politics by a will to art, by a consideration of the people qua work of art. . . . [A]esthetics can be understood in a Kantian sense re-examined perhaps by Foucault—as the system of *a priori* forms determining what presents itself to sense experience. It is a delimitation of spaces and times, of the visible and the invisible, of speech and noise, that simultaneously

determines the place and the stakes of politics as a form of experience. Politics revolves around what is seen and what can be said about it, around who has the ability to see and the talent to speak, around the properties of spaces and the possibilities of time." See Jacques Rancière, *The Politics of Aesthetics: The Distribution of the Sensible* (London: Continuum, 2004), 13.

35. Young, "Cultural Techniques and Logistical Media," np.

36. E.g., Lisa Gitelman, *Paper Knowledge: Toward a Media History of Documents*, Sign, Storage, Transmission, (Durham, NC: Duke University Press, 2014). See Jussi Parikka, *What Is Media Archaeology?* (Cambridge: Polity Press, 2012).

37. E.g., Alison Adam, "Lists," in *Software Studies: A Lexicon*, ed. Matthew Fuller (Cambridge: MIT Press, 2008), 174–78. Another source is Vismann, *Files*, 163.

38. Umberto Eco, *The Infinity of Lists*, trans. Alastair McEwen (New York: Rizzoli, 2009), 7.

39. Jorge Luis Borges, *Other Inquisitions, 1937–1952*, trans. Ruth L. C. Simms (University of Texas Press, 1964), 103.

40. When representing things other than numbers, these must first be encoded, i.e., transformed into a numerical representation that may be decoded into the original input with the correct approach. For example, the letters of the alphabet have a default order and may be encoded by counting a as 1, b as 10 (2), c as 11 (3) and so on. More complex objects, such as pictures, are often encoded recursively, e.g., with each pixel encoded separately as a series of three numbers (red, blue, green), and the entire picture is encoded as a series of pixels with added metadata that outlines the dimensions of the picture to be used when decoding.

41. Marcus Tullius Cicero, *Letters to Atticus*, trans. E. O. Winstedt (London: William Heinemann, 1913), 412.

42. See Calvert Watkins, *The American Heritage Dictionary of Indo-European Roots* (Boston: Houghton Mifflin, 1985), and Douglas R. Harper, "digit (n.)," in *Online Etymology Dictionary* (Tupelo, Mississippi, 2021).

43. The recent explosion in companies that deliver data-driven products has made data comparable in value to oil or even the very rays of the sun. See "The World's Most Valuable Resource Is No Longer Oil, but Data," *The Economist* (London), May 6, 2017, or Ludwig Siegele, "Are Data More Like Oil or Sunlight?" *The Economist* (London), February 20, 2020.

44. Robert Burch, "Charles Sanders Peirce," in *The Stanford Encyclopedia of Philosophy*, ed. Edward N. Zalta (Stanford, CA: Metaphysics Research Lab, Stanford University, 2022).

45. Igor Douven, "Abduction," in *The Stanford Encyclopedia of Philosophy*.

46. Charles S. Peirce, "Lowell Lectures on The Logic of Science; or Induction and Hypothesis: Lecture IX," in *Writings of Charles S. Peirce. Volume 1, 1857–1866: A Chronological Edition*, ed. Max H. Fisch (Bloomington: Indiana University Press, 1982).

47. Keith Green, "Deixis and Anaphora: Pragmatic Approaches," in *Encyclopedia of Language and Linguistics*, 2nd ed., ed. Keith Brown, 415–17 (Oxford: Elsevier, 2006), https://doi.org/10.1016/B0-08-044854-2/00328-X.

48. Maja Bak Herrie, "Unddragelsens kunst. Hermetiske objekter og deiksis uden kontekst i Gertrude Steins 'Tender Buttons,'" *Passage* 32, no. 77 (2017).

49. When speaking of a certain "deictic orientation" and even applying it as a sort of counterconcept to mathematical or technical understandings of digital phenomena, an author must have certain reservations. When I transfer this concept from its traditional use in the field of linguistics, I risk undermining its cogency and being imprecise. In linguistics, *deixis* is primarily used as a technical concept to describe certain terms, whereas the way I use it is metaphorical. I suggest using it to see broader or more general characteristics ascribed to deictic terms, that is, subjective experience on the one hand—a cognitive center of orientation, where language operates and assigns meaning to things and people—and, on the other hand, a "situational 'co-ordination'" of these people (I/you, us/them), places (here/there, this/that), and times (now/then, yesterday/today); e.g., Chris Baldick, "deixis," *The Oxford Dictionary of Literary Terms* (Oxford University Press, 2015); Green, "Deixis and Anaphora." Although this transfer from one discipline to another may cause certain problems, I nonetheless choose to draw on this theoretical resource, because in the same way that linguistic organizations of things, people, and places in time and space require mutual attention to contextual and situational conditions, and to the experimental orientation according to which the linguistic inquiry is received, digital representations and presentations also need to address how data sets and models relate to the surrounding world, and how its abstractions are received and interpreted.

50. Particularly in his later work, Maurice Merleau-Ponty criticizes the Cartesian legacy, which is also supplemented by a criticism of Jean-Paul Sartre's thinking. Sartre is wrong to separate subjects and objects according to a sharp distinction between the "in-itself" and "for-itself." Merleau-Ponty maintains that Sartre conceives subjectivity as holding being before itself, as a spectacle, and, hence, as not operating "from the middle of things." This contrasts with Merleau-Ponty's project, which specifically explores the *in-betweenness*, that is, the lived relations in which humans are embedded. See Maurice Merleau-Ponty, "La Nature ou le monde du silence: Pages d'introduction," in *Maurice Merleau-Ponty*, ed. Emmanuel de Saint Aubert (Paris: Hermann, 2008), 48. For an in-depth discussion of this idea of a middle, or "in-betweenness," in Merleau-Ponty's work, see A. S. Aurora Hoel and Annamaria Carusi, "Merleau-Ponty and the Measuring Body," *Theory, Culture & Society* 35, no. 1 (2018).

51. A central concept in Alfred North Whitehead's writings is the so-called *bifurcation of nature*, explained in *The Concept of Nature*, where he writes, "For natural philosophy everything perceived is in nature. We may not pick and choose. For us the red glow of the sunset should be as much part of nature as are the molecules

and electric waves by which men of science would explain the phenomenon." See Alfred North Whitehead, *The Concept of Nature* (Cambridge: Cambridge University Press, 1920), 29. In other words, this bifurcation of nature implies a way of thinking that divides the world in two: one that is composed of the fundamental constituents of the universe—invisible to the naked eye, but accessible with the right tools of measurement—and the other, which is constituted of what the mind must add to these basic building blocks of the world, for it to make sense; see Isabelle Stengers, *Thinking with Whitehead: A Free and Wild Creation of Concepts* (Cambridge, MA: Harvard University Press, 2011), xii. According to Whitehead, although the natural sciences start with the first—the molecules and the electrical waves, that is, from below—the social sciences and humanities start from above, from the redness of the sunset and its warmth on our bodies. However, the problem with this kind of thinking is that it challenges basic philosophical questions, e.g., about the status of the mind or the nature of subjective experience. "If nature really is bifurcated, no living organism would be possible, since being an organism means being the sort of thing whose primary and secondary qualities—if they exist—are endlessly blurred. Since we are organisms surrounded by other organisms, nature has *not* bifurcated," Bruno Latour writes in his foreword to Stenger's book, see Stengers, *Thinking with Whitehead*, xiii. Latour mentions this division again in his translation of Whitehead's terms into his own pair of concepts, *matters of fact* and *matters of concern*. With these concepts, he argues that we should approach *things* in their double meaning, from the Old English and German *Ding*, as both an object out there and a concern: "Icelanders boast of having the oldest Parliament, which they call Althing, and you can still visit in many Scandinavian countries assembly places that are designated by the word *Ding* or *Thing*. Now, is this not extraordinary that the banal term we use for designating what is out there, unquestionably, a thing, what lies out of any dispute, out of language, is also the oldest word we all have used to designate the oldest of the sites in which our ancestors did their dealing and tried to settle their disputes? A thing is, in one sense, an object out there and, in another sense, an *issue* very much *in* there, at any rate, a *gathering*. [T]he same word *thing* designates matters of fact and matters of concern," see Bruno Latour, "Why Has Critique Run Out of Steam? From Matters of Fact to Matters of Concern," *Critical Inquiry* 30, no. 2 (2004): 233.

52. Mieke Bal, *Travelling Concepts in the Humanities* (University of Toronto Press, 2002), 45.

53. See Juliane Rebentisch, *Theorien der Gegenwartskunst zur Einführung* (Hamburg: Junius, 2013). For an in-depth discussion of the capability of artworks and exhibitions to produce "reflexive transformations" of otherwise nonartistic issues, see Jacob Lund, "Exhibition as Reflexive Transformation," *OBOE Journal* 3, no. 1 (2022): i–x.

54. Although using the concept of knowledge might contribute to a problematic prioritization of discursive over aesthetic or visual approaches, Tom Holert

in his 2020 book *Knowledge Beside Itself: Contemporary Art's Epistemic Politics* (London: Sternberg Press) nonetheless argues that by engaging in the "the activation, reconstruction, resurrection, recomposition, and invention of ways of knowing and modes of thought that are irreducible to Western rationalism and cognitivism, the aesthetic might be regained as the reservoir and repertoire of a cognition that is based in bodily, sensations, in affect, in empathy" (61). In alignment with this perspective, I use a similar conception of "knowledge production" within this book, viewing knowledge as an epistemic activity encompassing language use, thinking, learning, and archiving, or "social organizing" (45).

55. For an in-depth discussion of this "operative" approach to images, see A. S. Aurora Hoel, "Operative Images. Inroads to a New Paradigm of Media Theory," in *Image–Action–Space: Situating the Screen in Visual Practice,* ed. Luisa Feiersinger, Kathrin Friedrich, and Moritz Queisner, 11–28. (Berlin, Boston: De Gruyter, 2018). These ideas draw in part on Harun Farocki's conception of active images, see Harun Farocki, "Phantom Images," *Public* 29 (2004), https://public.journals .yorku.ca/index.php/public/article/view/30354. See also Sybille Krämer, "Operative Bildlichkeit. Von der 'Grammatologie' zu einer 'Diagrammatologie'? Reflexionen über erkennendes Sehen" (Bielefeld: Transcript Verlag, 2015).

56. See Chiara Ambrosio, "Composite Photographs and the Quest for Generality: Themes from Peirce and Galton," *Critical inquiry* 42, no. 3 (2016) and Lorraine Daston and Peter Galison, "The Image of Objectivity," *Representations*, no. 40 (1992): 98–117.

57. For a discussion of the idea of "data-driven" epistemologies, see Kitchin, "Big Data, New Epistemologies and Paradigm Shifts." This approach to knowledge production has received extensive criticism, e.g., danah boyd and Kate Crawford, "Six Provocations for Big Data" (paper presented at the A Decade in Internet Time: Symposium on the Dynamics of the Internet and Society, 2011).

58. E.g., Luciana Parisi, "Critical Computation: Digital Automata and General Artificial Thinking," *Theory, Culture & Society* 36, no. 2 (2019). See also Pedro Domingos, *The Master Algorithm: How the Quest for the Ultimate Learning Machine Will Remake Our World* (New York: Basic Books, 2015); David Bollier and Charles M. Firestone, *The Promise and Peril of Big Data* (Washington, DC: Aspen Institute, Communications and Society Program, 2010); and Luciano Floridi, "Big Data and Their Epistemological Challenge," *Philosophy & Technology* 25, no. 4 (2012).

59. See Simon Aagaard Enni and Maja Bak Herrie, "Turning Biases into Hypotheses through Method: A Logic of Scientific Discovery for Machine Learning," *Big Data & Society* 8, no. 1 (2021). For an in-depth discussion of the potential problems of using of machine learning in scientific knowledge production, see Simon Aagaard Enni, "Deliberation and Dissemination in Machine Learning" (Aarhus: Computer Science, Aarhus University, 2021).

60. See Lisa Gitelman, *"Raw Data" Is an Oxymoron* (Cambridge, MA: MIT Press, 2013).

61. Charles Sanders Peirce, *The Collected Papers of Charles Sanders Peirce*, electronic ed., ed. Charles Hartshorne, Paul Weiss, and Arthur W. Burks, vol. 5, *Pragmatism and Pragmaticism* (Cambridge, MA: Belknap Press of Harvard University Press, 1994). See also Hoel, "Lines of Sight."

Chapter 1

1. Lorraine Daston, "Why Statistics Tend Not Only to Describe the World but to Change It," *London Review of Books* 22, no. 8 (2000).

2. Parts of the analysis presented in this chapter has previously been published in Maja Bak Herrie, "Tracing the Outlier: Digital Objects and Algorithmic Sorting in Rossella Biscotti's *Other*," *Journal of Aesthetics & Culture* 11, no. 1 (2018).

3. From the middle of the eighteenth century, the adoption of machines, typically powered by water wheels and later, steam engines, meant that many skilled textile workers lost their employment, leading to protest movements such as the Luddites. Although new, less-skilled jobs were created, the poor working conditions in the textile mills and the notoriously low wages continued to impoverish the workers; e.g., Mark Cartwright, "The Textile Industry in the British Industrial Revolution," in *World History Encyclopedia*. Last modified March 1, 2023, https://www.worldhistory.org/article/2183/the-textile-industry-in-the-british-industrial-rev/ or Melinda Watt, "Nineteenth-Century European Textile Production," in *Heilbrunn Timeline of Art History* (New York: The Metropolitan Museum of Art, 2004).

4. According to the *Oxford English Dictionary*, the term *computer* was first used at the beginning of the seventeenth century and simply referred to "one who computes," that is, a person who executed mathematical calculations before the electronic computer was available. Alan Turing described this occupation in the following way: "The human computer is supposed to be following fixed rules; he has no authority to deviate from them in any detail," see A. M. Turing, "I.—Computing Machinery and Intelligence," *Mind* 59, no. 236 (1950). Often, however, women, rather than men, did this formalized calculation. For instance, women were employed by the United States Signal Corps to make calculated predictions about the weather, and by Harvard University to process astronomical data under the name of the "Harvard Computers," see David Alan Grier, *When Computers Were Human* (Princeton, NJ: Princeton University Press, 2013). Later, female "computers" were also hired by NASA, as depicted in the 2016 Hollywood drama, *Hidden Figures*. See also M. L. Shetterly's 2016 book, *Hidden Figures: The American Dream and the Untold Story of the Black Women Mathematicians Who Helped Win the Space Race* (New York: William Morrow).

5. These systems' capacity to process large-scale data sets to not merely describe and represent worlds, but also to *generate* new ones, has developed in a complex infrastructure made up of technologies and epistemologies, see Luciana Parisi, "Recursive Philosophy and Negative Machines," *Critical Inquiry* 48, no. 2 (2022).

However, this development is haunted by the histories of colonialism, race, and demography in which statistics had its modern origins. For more on the connection between colonialism and data extraction see, e.g., Nick Couldry and Ulises A. Mejias, "Data Colonialism: Rethinking Big Data's Relation to the Contemporary Subject," *Television & New Media* 20, no. 4 (2019). For more on how policing, biometrics, and data reifications relate to categories of race and blackness, see Simone Browne, *Dark Matters: On the Surveillance of Blackness* (Durham, NC, and London: Duke University Press, 2015).

6. E.g., Catherine D'Ignazio and Lauren F. Klein, "Feminist Data Visualization" (IEEE VIS Conference: Workshop on Visualization for the Digital Humanities 2015). See also Tamara Munzner and Éamonn Maguire, *Visualization Analysis & Design*, A K Peters Visualization Series (Boca Raton, FL: CRC Press, 2015).

7. E.g., Maja Bak Herrie, "Topological Thinking-Machines," in *Postheroic Supersession*, ed. Jørgen Michaelsen and Lucas Haberkorn (Silkeborg: Museum Jorn, 2022).

8. V. Barnett and T. Lewis, *Outliers in Statistical Data*, 3rd ed. (Hoboken, NJ: Wiley, 1994), 3.

9. Several statistical tests rely on the so-called *assumption of normality*, which means that the data analyzed follow a normal distribution. Some of the commonly used statistical tests that assume normality include the student's t-test, analysis of variance (ANOVA), linear regression, and the chi-squared test.

10. Frank J. Anscombe, "Rejection of Outliers," *Technometrics* 2, no. 2 (1960): 125.

11. Alain Desrosières, *The Politics of Large Numbers: A History of Statistical Reasoning* (Cambridge, MA: Harvard University Press, 1998), 147–50.

12. Anscombe, "Rejection of Outliers," 125.

13. Anscombe.

14. Arthur Zimek and Peter Filzmoser, "There and Back Again: Outlier Detection between Statistical Reasoning and Data Mining Algorithms," *Wiley Interdisciplinary Reviews. Data Mining and Knowledge Discovery* 8, no. 6 (2018).

15. Parts of this close reading of Rossella Biscotti's artistic project *Other* has been previously published in the article, "Tracing the Outlier: Digital Objects and Algorithmic Sorting in Rossella Biscotti's Other," *Journal of Aesthetics & Culture* 11, no. 1 (2019).

16. Rossella Biscotti, *Rossella Biscotti: 10 x 10*, trans. Barbara Hess and Michael Wolfson, ed. Magdalena Holzhey (Krefeld: Revolver Publishing, 2015).

17. To paraphrase the words of Ada Augusta King, Countess of Lovelace, who played an important role in the work of English mathematician Charles Babbage and who, in her translation of his memoirs noted that "we may say most aptly that the Analytical Engine weaves algebraic patterns just as the Jacquard loom weaves flowers and leaves." See Luigi Federico Menabrea and Charles Babbage, *Sketch of the Analytical Engine Invented by Charles Babbage. With Notes by the Translator*

Ada Augusta King, Countess Lovelace, trans. Augusta Ada King Lovelace (Richard & John E. Taylor, 1843).

18. Lisa Gitelman, *"Raw Data" Is an Oxymoron*, Infrastructures Series (Cambridge, MA: MIT Press, 2013), 9.

19. Wendy Hui Kyong Chun, *Updating to Remain the Same: Habitual New Media* (Cambridge, MA: MIT Press, 2016).

20. Giorgio Agamben, *What Is an Apparatus? And Other Essays* (Stanford, CA: Stanford University Press, 2009), 14.

21. Biscotti, *Rossella Biscotti: 10 x 10*, 13.

22. Biscotti, 17–18.

23. Biscotti, 15.

24. Markus Brüderlin, "Zur Ausstellung. Die Geburt der Abstraktion aus dem Geiste des Textilen und die Eroberung des Stoff-Raumes," in *Kunst & Textil. Stoff als Material und Idee in der Moderne von Klimt bis heute*, ed. Markus Brüderlin, 34–41 (Ostfildern: Kunstmuseum Wolfsburg and Staatsgalerie Stuttgart, 2013).

25. Biscotti, *Rossella Biscotti: 10 x 10*, 17.

26. Rosalind Krauss, "Grids," *October* 9 (1979): 52.

27. Bodil Marie Stavning Thomsen, "Signaletic, Haptic and Real-Time Material," *Journal of Aesthetics & Culture* 4, no. 1 (2012).

28. Julian Heynen, *Ein Ort der denkt: Haus Lange und Haus Esters von Ludwig Mies van der Rohe. Moderne Architektur und Gegenwartskunst* (Krefeld: Krefelder Kunstmuseen, 2000), 39.

29. Dominic Lopes, *A Philosophy of Computer Art* (London: Routledge, 2010), 8–14.

30. Here, the Greek word, *aisthêsis*, refers to an understanding of aesthetics that pertains to sensation and perception (in contrast to intellectual concepts or rational knowledge, that is, *noesis*).

31. Biscotti, *Rossella Biscotti: 10 x 10*.

32. Biscotti, 17.

33. Lopes, *A Philosophy of Computer Art*, 8–14; Lotte Philipsen, "Computerens plads i samtidskunsten: Fra et teknisk til et æstetisk paradigme," in *Cybermuseologi: Kunst, museer og formidling i et digitalt perspektiv*, ed. Ane Hejlskov Larsen, Rune Gade, and André Wang Hansen (Aarhus: Aarhus Universitetsforlag, 2015), 71–73; Christiane Paul, *A Companion to Digital Art*, ed. Dana Arnold (Hoboken, NJ: Wiley, 2016), 2.

34. Meredith Hoy, *From Point to Pixel: A Genealogy of Digital Aesthetics* (Dartmouth College Press, 2017), 8.

35. Lopes, *A Philosophy of Computer Art*.

36. Lopes, 8.

37. Biscotti, *Rossella Biscotti: 10 x 10*, 15.

38. Rossella Biscotti, Julia Reist, and Patrick Allo, "Other," *Online Journal of Contour Biennale* (2017), http://hearings.contour8.be/2017/05/15/other/.

39. Biscotti, Reist, and Allo, "Other."

40. According to an article authored by Rossella Biscotti, Julia Reist, and Patrick Allo, the project's point of departure was precisely the traditional structures of Mies van der Rohe's building. They write: "Haus Esters was built in a classical family structure with a family room, a man's room, a woman's room, and a children's room. With this rather traditional idea of how a family is composed as a starting point, we looked at the definition of households within the population census of Brussels. We examined the lives and relationships within different households included in this data set and ultimately replotted this demographic analysis into the form of woven textiles," see Biscotti, Reist, and Allo, "Other." Also, Adam Kleinman's essay "Herrenzimmer, Damenzimmer, Kinderzimmer" should be mentioned in this regard, as it, too, describes the traditional organization of the rooms in the villa. See Biscotti, *Rossella Biscotti: 10 x 10*, 29–38.

41. Watt, "Nineteenth-Century European Textile Production."

42. Sybille Krämer and Horst Bredekamp, "Culture, Technology, Cultural Techniques—Moving Beyond Text," *Theory, Culture & Society* 30, no. 6 (2013): 25.

43. The algorithm stands as one of the foundational concepts in computer science, playing a pivotal role in shaping the modern computer. Its historical and conceptual impact is profound, often encapsulated by the equation "Algorithm = Logic + Control." Andrew Goffey, in his lexical definition in *Software Studies*, describes it as the unifying concept for all activities undertaken by computer scientists. Goffey characterizes the algorithm as a "description of the method by which a task is to be accomplished," emphasizing its fundamental nature in the realm of computer science. See Andrew Goffey, "Algorithm," in *Software Studies: A Lexicon*, ed. Matthew Fuller (Cambridge, MA: MIT Press, 2008). As a rule, algorithms are not necessarily digital, but may be executed by hand, as Tarleton Gillespie writes: "We might think of computers . . . fundamentally as algorithm machines—designed to store and read data, apply mathematical procedures to it in a controlled fashion, and offer new information as the output. But these are procedures that could conceivably be done by hand—and in fact were." See Tarleton Gillespie et al., *Media Technologies: Essays on Communication, Materiality, and Society*, Inside Technology (Cambridge, MA: MIT Press, 2014), 167.

44. Alfred Barlow, *The History and Principles of Weaving by Hand and by Power* (University of Michigan: Low, Marston, Searle, & Rivington, 1878), 129–39.

45. Barlow, *The History and Principles of Weaving by Hand and by Power*, 141.

46. Today, Jacquard is often portrayed as the inventor of the automatic loom and of the subsequent mechanization of weaving that changed the textile industry. Yet, this linear progression from the Jacquard machine, to Herman Hollerith's tabulation machine, to modern data processing is not unproblematic. As briefly described above, the first system was originally developed for the manual drawloom (and not the so-called power loom) by the Lyon workers Bouchon and Falcon in the 1720s, and it took many attempts and iterations to run the complex

patterns that are familiar to us from the later power looms, creatively marketed by entrepreneurs such as Jacquard.

47. Thomas Palmelund Johansen, "Automaternes sociale liv—væven, computeren og menneskene," *Baggrund, June 5, 2016*, https://baggrund.com/2016/06/05/automaternes-sociale-liv-vaeven-computeren-og-menneskene/.

48. Biscotti, *Rossella Biscotti: 10 x 10*, 43.

49. E.g., IBM's series, "Icons of Progress," where "The Punched Card Tabulator" tells a story about Jacquard's inventions and how he developed the automatic loom, which would later be succeeded by the English mathematician Charles Babbage's systematic difference engine. This engine's conceptual way of operating would then come in the hands of Herman Hollerith, who would develop a tabulating machine that "opened the world's eyes to the very idea of data processing," an idea that, over time, would lay "the foundation for IBM," that is, "one of the few major corporate success stories of the Great Depression" of the 1930s, and also "launched the company on its path to becoming a computing giant." We should not forget IBM's historical use of the tabulating machine, first to count and control Asian immigrants and other so-called undesirables, (Biscotti, *Rossella Biscotti: 10 x 10*, 41), and subsequently, by the German census of 1933, which documented Jews, Romani people, and other stigmatized ethnic groups, and finally, to register the number and location of said minorities to collect them in concentration camps. Edwin Black has studied this particularly dark chapter of the history of data processing in his very informative book; see Edwin Black, *IBM and the Holocaust: The Strategic Alliance Between Nazi Germany and America's Most Powerful Corporation* (Washington, DC: Dialog Press, 2009).

50. Daston, "Why Statistics Tend Not Only to Describe the World but to Change It."

51. Krämer and Bredekamp, "Culture, Technology, Cultural Techniques—Moving Beyond Text," 24.

52. In their 2018 book, *The Metainterface*, Christian Ulrik Andersen and Søren Bro Pold stress the importance of a new understanding of the metainterface, which describes the "situation where the computer's interface seemingly both becomes omnipresent and invisible, and where it at once is embedded in everyday objects and characterized by hidden exchanges of information between objects." See Christian Ulrik Andersen and Soren Bro Pold, *The Metainterface: The Art of Platforms, Cities, and Clouds* (Cambridge, MA: MIT Press, 2018), 5. Although the interface may not always be visible or "available," it is nevertheless pervasive in Andersen and Pold's conceptualization: multifaceted and powerful, and an important part of our digital milieu and culture, the metainterface is far from just formally algorithmic.

53. Krämer and Bredekamp, "Culture, Technology, Cultural Techniques—Moving Beyond Text," 24.

54. Krämer and Bredekamp, 24.

55. Cornelia Vismann, "Cultural Techniques and Sovereignty," *Theory, Culture & Society* 30, no. 6 (2013): 83.

56. Krämer and Bredekamp, "Culture, Technology, Cultural Techniques—Moving Beyond Text," 26.

57. Friedrich A. Kittler and John Johnston, *Literature, Media, Information Systems: Essays*, Critical Voices in Art, Theory and Culture (Amsterdam: Gordon & Breach, 1997), x.

58. Sybille Krämer, *Symbolische Maschinen: Die Idee der Formalisierung in geschichtlichem Abriss* (Darmstadt: Wissenschaftliche Buchgesellschaft, 1988).

59. Cornelia Vismann, *Files: Law and Media Technology*, trans. Geoffrey Winthrop-Young, Meridian, Crossing Aesthetics (Stanford, CA: Stanford University Press, 2008), 83.

60. Liam Cole Young, "Cultural Techniques and Logistical Media: Tuning German and Anglo-American Media Studies," *M/C journal* 18, no. 2 (2015).

61. Biscotti, Reist, and Allo, "Other."

62. Geoffrey C. Bowker and Susan Leigh Star, *Sorting Things Out: Classification and Its Consequences*, 4th ed., Inside Technology (Cambridge, MA: MIT Press, 2002), 10.

63. Bowker and Star, *Sorting Things Out: Classification and Its Consequences*, 11.

64. Biscotti, Reist, and Allo, "Other."

65. Luciana Parisi, "Negative Optics in Vision Machines," *AI & Society* 36, no. 4 (2021).

66. E.g., Denise Ferreira de Silva's analysis of the relations between formlessness and value and the Western idea that only through formal representation in the ethical regime can a subject exist and *matter*. "Blackness," as an epistemic category, is debarred from such existence, as it is rendered formless. See Denise Ferreira da Silva, "1 (life) ÷ 0 (blackness) = ∞ − ∞ or ∞ / ∞: On Matter beyond the Equation of Value," *e-flux* 79, February (2017).

67. Benjamin Peirce, "Criterion for the Rejection of Doubtful Observations," *Astronomical Journal* 2 (1852). For an introduction to Benjamin Peirce's criterion for rejection, see Desrosières, *The Politics of Large Numbers*, 220.

68. Anscombe, "Rejection of Outliers," 126.

69. V. Barnett and T. Lewis define the outlier as "an observation (or subset of observations) which appears to be inconsistent with the remainder of that set of data." Barnett and Lewis, *Outliers in Statistical Data*, 7. Similarly, A. Zimek and P. Filzmoser describe outliers as "data objects that do not fit well to the general data distribution." Zimek and Filzmoser, "There and Back Again," 1.

70. Today, statistical methods are supplemented by more geometrical tools, resulting, for example, in the pervasiveness of linear algebra in outlier rejection; see Zimek and Filzmoser, "There and Back Again."

71. See Victoria Hodge and Jim Austin, "A Survey of Outlier Detection Methodologies," *Artificial intelligence Review* 22, no. 2 (2004). Another reference is Zimek and Filzmoser's "There and Back Again," 2.

72. Zimek and Filzmoser, "There and Back Again," 4–5.

73. The "wastebasket" group is used in taxonomy to describe a group of organisms that do not fit any of the existing taxons. See Stephen Jay Gould, "Treasures in a Taxonomic Wastebasket," *Natural History* 94 (1985), 22–33.

74. Donna Haraway, "Situated Knowledges: The Science Question in Feminism and the Privilege of Partial Perspective," *Feminist Studies* 14, no. 3 (1988).

75. Liv Hausken, *Thinking Media Aesthetics: Media Studies, Film Studies and the Arts* (Frankfurt am Main: Peter Lang International Academic Publishing Group, 2013).

76. Kittler and Johnston, *Literature, Media, Information Systems*, x.

77. Geoffrey C. Bowker, *Memory Practices in the Sciences*, Inside Technology (Cambridge, MA: MIT Press, 2005), 183–84.

78. Daston, "Why Statistics Tend Not Only to Describe the World but to Change It."

79. Mieke Bal, *Travelling Concepts in the Humanities: A Rough Guide*, Green College Lecture Series (Toronto: University of Toronto Press, 2002), 22.

Chapter 2

1. Jorge Luis Borges, "Funes the Memorious," in *Labyrinths: Selected Stories and Other Writings*, ed. Donald Alfred Yates and James E. Irby (London: Penguin, 1970), 65.

2. Elyce Rae Helford, "Language and Memory in Borges' 'Funes, The Memorious,'" *Iowa Journal of Literary Studies* 9, no. 1 (1988): 15.

3. Ferdinand de Saussure, *Course in General Linguistics*, ed. Charles Bally, Albert Sechehaye, and Albert Reidlinger, trans. Wade Baskin (New York: Philosophical Library, 1959), 122–23.

4. I was inspired to relate Borges's protagonist, Funes, and data processing by the chapter on aggregation in S. M. Stigler's 2016 book, *The Seven Pillars of Statistical Wisdom* (Cambridge, MA: Harvard University Press, 2016).

5. The word *data* has its etymological roots in the Latin *datum*, which simply means the "given"; see Lisa Gitelman and Virginia Jackson's introduction to *"Raw Data" Is an Oxymoron*, Infrastructures Series (Cambridge, MA: MIT Press, 2013), 7.

6. Michael A. Stoto and John D. Emerson, "Power Transformations for Data Analysis," *Sociological Methodology* 14 (1983): 127.

7. Lev Manovich, *The Language of New Media*, Leonardo (Cambridge, MA: MIT Press, 2001), 224.

8. Gitelman, *"Raw Data" Is an Oxymoron*, 3 (emphasis is mine).

9. Gitelman, 2.

10. Geoffrey C. Bowker, *Memory Practices in the Sciences*, Inside Technology (Cambridge, MA: MIT Press, 2005), 183–84; Gitelman, *"Raw Data" Is an Oxymoron*, 2.

11. For an in-depth discussion of data selection, see Simon Aagaard Enni and

Maja Bak Herrie's "Turning Biases into Hypotheses through Method: A Logic of Scientific Discovery for Machine Learning," *Big Data & Society* 8, no. 1 (2021).

12. Stoto and Emerson, "Power Transformations for Data Analysis," 127.

13. Although the visual presentation of the portraits is rendered on a flat surface (i.e., the screen), the facial models are represented internally as three-dimensional, geometric meshes within the system. For an in-depth media-theory and cultural-historical discussion of flatness and the operation of "flattening" from the perspective of the humanities, see Sybille Krämer, "The 'Cultural Technique of Flattening.' An Essay Introducing and at the Same Time Revising an Idea," *Metode* 1 (2023).

14. Stigler, *The Seven Pillars of Statistical Wisdom*, 13.

15. I will not go into detail related to these types of presentations in this chapter. For a discussion of these methods, see Stigler's chapter on aggregation in *The Seven Pillars of Statistical Wisdom*.

16. This example is also emphasized by Stigler in his description of the average, see Stigler, *The Seven Pillars of Statistical Wisdom*, 131–32.

17. Stigler, *The Seven Pillars of Statistical Wisdom*, 14.

18. Stigler, 33.

19. Alain Desrosières, *The Politics of Large Numbers: A History of Statistical Reasoning* (Cambridge, MA: Harvard University Press, 1998), 67.

20. A product of his time, Quetelet only included men in his demographic studies, whereas women and other genders "escaped" his reductions, see Stephen M. Stigler, *The History of Statistics: The Measurement of Uncertainty before 1900* (Cambridge, MA: Belknap Press of Harvard University Press, 1986), 169–74.

21. Peter L. Bernstein, *Against the Gods: The Remarkable Story of Risk* (Chichester: Wiley, 1998), 159.

22. Lambert Adolphe Jacques Quetelet, *A Treatise on Man and the Development of His Faculties*, ed. T. Smibert, trans. R. Knox (Edinburgh: William and Robert Chambers, 1842), 100.

23. Bernstein, *Against the Gods*, 159.

24. The quote from Cournot's work *Exposition de la théorie des chances et des probabilités* (1843) is included in Stigler's English translation, see Stigler, *The History of Statistics*, 195–96.

25. Bernstein, *Against the Gods*, 159–60.

26. Claude Bernard, *An Introduction to the Study of Experimental Medicine*, trans. H. C. Green (New York: Dover, [1927] 1957) 134–35.

27. Quetelet, *A Treatise on Man and the Development of His Faculties*, 6.

28. Quetelet, 118.

29. Bernstein, *Against the Gods*, 162.

30. Ramon Amaro, "As If," *e-flux Architecture* (2019): 5.

31. Chiara Ambrosio, "Composite Photographs and the Quest for Generality: Themes from Peirce and Galton," *Critical Inquiry* 42, no. 3 (2016): 551.

32. Stigler, *The Seven Pillars of Statistical Wisdom*, 35.

33. Francis Galton, *Inquiries into the Human Faculty and Its Development* (London: Macmillan and Co., 1883), 350.

34. Geoffrey C. Bowker et al., "Toward Information Infrastructure Studies: Ways of Knowing in a Networked Environment," in *International Handbook of Internet Research*, ed. Jeremy Hunsinger, Lisbeth Klastrup, and Matthew Allen (Dordrecht: Springer Netherlands, 2010), 111.

35. E.g., Taina Bucher, *If... Then: Algorithmic Power and Politics* (New York: Oxford University Press, 2018); Louise Amoore, "Data Derivatives: On the Emergence of a Security Risk Calculus for Our Times," *Theory, Culture & Society* 28, no. 6 (2011).

36. Viktor Mayer-Schönberger and Kenneth Cukier, *Big Data: A Revolution That Will Transform How We Live, Work, and Think* (Boston: Mariner Books, 2014), 6.

37. Chris Anderson's controversial article, "The End of Theory: The Data Deluge Makes the Scientific Method Obsolete," published in *Wired* in 2008, presents perhaps the most emphatic example of this idea. It has been extensively debated and criticized, however; see, Alexander Campolo and Kate Crawford, "Enchanted Determinism: Power without Responsibility in Artificial Intelligence," *Engaging Science, Technology, and Society* 6, no. 1 (2020); Rob Kitchin, "Big Data, New Epistemologies and Paradigm Shifts," *Big Data & Society* 1, no. 1 (2014); danah boyd and Kate Crawford, "Critical Questions for Big Data: Provocations for a Cultural, Technological, and Scholarly Phenomenon," *Information, Communication & Society* 15, no. 5 (2012).

38. Gitelman, *"Raw Data" Is an Oxymoron*, 1.

39. Antoinette Rouvroy and Thomas Berns, "Algorithmic Governmentality and Prospects of Emancipation: Disparateness as a Precondition for Individuation Through Relationships?," *Réseaux* 177, no. 1 (2013).

40. Gilles Deleuze, "Postscript on Control Societies," in *Negotiations, 1972–1990* (New York: Columbia University Press, 1995).

41. Orit Halpern et al., "Surplus Data: An Introduction," *Critical Inquiry* 48, no. 2 (2022).

42. Gitelman, *"Raw Data" Is an Oxymoron*, 8.

43. Gitelman, 8.

44. Adam Broomberg and Oliver Chanarin, *Spirit Is a Bone* (London: Mack, 2016), 238.

45. Artworks' political nature is noticeable on (at least) three levels. First, there is the use of software such as *FaceControl 3-D*, which relates directly to the Russian company, Vocord, which provided the equipment for Broomberg and Chanarin's project, but who also supply surveillance solutions for places that include subway stations and transport hubs in Ryazan, and the Omsk Arena, where it is used to keep unruly hockey fans out of the stadium; see, Laura Mallonee, "Creepy Por-

traits Made with Even Creepier Surveillance Tech," *Wired* February 16, 2016. Secondly, some participants, such as Pussy Riot member Yekaterina Samutsevich and writer Lev Rubinstein, are political voices in Russia. Lastly, the artists themselves are involved in the political interpretation of their project. They acknowledge the political agenda behind their investigation and capture faces in various locations. For instance, the *Guardian* quoted them saying that "[t]he aim [of the project] is to take shots of people passing through places like border crossings, railway stations, sports halls, even cinemas . . . It is eerie and sinister: it captures the shape of a face in a split second, from multiple angles, using various lenses. It then constructs a 3D model of the head that can be closely analysed and stored for future reference"; see S. O'Hagan: "Negative Humanity: The Birth of the Digital Death Mask," *The Guardian*, February 5, 2016. I will not go further into this specific political level of the project, but instead examine the specific ways in which this digital aggregation logic becomes evident. Scholars such as Lisa Gitelman and Lorraine Daston inspired me to explore how this digital aggregation logic is manifested and to focus on its concrete manifestations rather than normative judgments; see Peter Galison and Lorraine Daston, *Objectivity* (New York: Zone Books, 2007), 51 and Gitelman, *"Raw Data" Is an Oxymoron*, 4.

46. E.g., T. Matzner, "Why Privacy Is Not Enough Privacy in the Context of 'Ubiquitous Computing' and 'Big Data,'" *Journal of Information, Communication and Ethics in Society* 12, no. 2 (2014), or S. D. Esposti, "When Big Data Meets Dataveillance: The Hidden Side of Analytics," *Surveillance & Society* 12, no. 2 (2014).

47. Gemma Galdon Clavell, "Big Data," in *The SAGE Encyclopedia of Surveillance, Security, and Privacy*, ed. Gemma Galdon Clavell, 99–101. Thousand Oaks, CA: SAGE Publications, 2018, https://doi.org/10.4135/9781483359922.

48. Clavell, "Big Data," 100.

49. Esposti, "When Big Data Meets Dataveillance."

50. Several researchers have pointed out that the field of surveillance studies has been dominated by concepts of the panopticon, Big Brother, and George Orwell's famous work, *1984*. See, K. D. Haggerty and R. V. Ericson, "The Surveillant Assemblage," *British Journal of Sociology* 51, no. 4 (2000), and D. Lyon: *Surveillance Studies: An Overview* (Cambridge: Polity Press, 2007). However, in recent years, the emergence of new technologies and actors have been emphasized as crucial factors in the development of a more multifaceted phenomenon. See, David Lyon, "Reflections on Forty Years of 'Surveillance Studies,'" *Surveillance & Society* 20, no. 4 (2022): 353–56.

51. The conversation, titled "The Bone Cannot Lie," between Eyal Weizman and Adam Broomberg and Oliver Chanarin was conducted on September 27, 2015, and a transcribed version can be found in the book as well as the associated website, http://www.broombergchanarin.com/article-the-bone.

52. The death mask dates to 2400 BCE, when the Egyptians used it to create portraits of the faces of the deceased. First, a negative imprint was taken in mate-

rials such as plaster or wax, which could then be used to create a positive portrait in the desired material. Later, death masks were made of numerous individuals, including writers, artists, politicians, and royalty. See *Oxford English Dictionary*, under the entry "death."

53. N. Dagnes, E. Vezzetti, F. Marcolin, et al., "Occlusion Detection and Restoration Techniques for 3D Face Recognition: A Literature Review," *Machine Vision and Applications*, 29, no. 5 (2018), 790.

54. Dagnes, Vezzetti, Marcolin et al., "Occlusion Detection and Restoration Techniques for 3D Face Recognition," 790.

55. "The Bone Cannot Lie," http://www.broombergchanarin.com/article-the -bone.

56. Broomberg and Chanarin.

57. E.g., Harun Farocki, "Phantom Images," *Public* 29 (2004); A. S. Aurora Hoel and Frank Lindseth, "Differential Interventions: Images as Operative Tools," *Photomediations: A Reader*, ed. Kamila Kuc and Joanna Zylinska (Open Humanities Press, 2016); Ingrid Hoelzl and Remi Marie, *Softimage: Towards a New Theory of the Digital Image* (Bristol, UK: Intellect Books, 2015); Sybille Krämer, "Operative Bildlichkeit: Von der 'Grammatologie' zu einer 'Diagrammatologie'? Reflexionen über erkennendes 'Sehen,'" *Logik des Bildlichen: Zur Kritik der ikonischen Vernunft*, ed. M. Hessler og D. Mersch (Bielefeld: Transcript, 2009); or Trevor Paglen, "Invisible Images (Your Pictures Are Looking at You)," *New Inquiry*, December 8, 2016, https: //thenewinquiry.com/invisible-images-your-pictures-are-looking-at-you/.

58. Hoel and Lindseth, "Differential Interventions," 177–78.

59. Farocki, "Phantom Images," 17.

60. Jussi Parikka, *Operational Images: From the Visual to the Invisual* (Minneapolis: University of Minnesota Press, 2023).

61. Paglen, "Invisible Images (Your Pictures Are Looking at You)," https:// thenewinquiry.com/invisible-images-your-pictures-are-looking-at-you/.

62. Hoel and Lindseth, "Differential Interventions," 178.

63. William Uricchio, "The Algorithmic Turn: Photosynth, Augmented Reality and the Changing Implications of the Image," *Visual Studies* 26, no. 1 (2011): 33.

64. W.J.T. Mitchell, *Picture Theory: Essays on Verbal and Visual Representation* (Chicago: University of Chicago Press, 1995).

65. Mitchell, *Picture Theory*, 36.

66. Mitchell, 57.

67. See more regarding machine vision images and Mitchell's concept of metapictures in Fabian Offert and Peter Bell's, "Perceptual Bias and Technical Metapictures: Critical Machine Vision as a Humanities Challenge," *AI & Society* 36, no. 8 (2021): 1133–44.

68. Paglen, "Invisible Images (Your Pictures Are Looking at You)."

69. Johanna Zylinska, *Nonhuman Photography* (Cambridge, MA: MIT Press, 2017).

70. Many artists work with(in) the human body and explore the intimate imagery of real-time surgical procedures. An example of this is Yuri Ancarani's piece, *Da Vinci* (2012).

71. An example of an artistic investigation of the role of thermal imagery in the refugee and migration situation now affecting Europe, the Middle East, and North Africa is Richard Mosse's project, *Heat Maps* (2016). Mosse utilizes extreme tele-military-grade thermographic cameras to map refugee camps and other staging sites.

72. An example is Abelardo Gil-Fournier's piece, *Landscape Prediction: An Earthology of Moving Landforms* (2018), which employs commercial satellite imagery to explore the cinematic character of active landforms such as river thalwegs, drifting glaciers, or crawling dunes.

73. Timothy Morton, "Poisoned Ground: Art and Philosophy in the Time of Hyperobjects," *Symploke* 21, no. 1 (2013): 40.

74. With regard to photographic portraiture as a system of representation, the American theorist and artist Allan Sekula may be identified as an important exponent of this position. To him, portrait photography may be either *honorific* or *repressive*, depending on the person portrayed. It may extend and popularize—particularly the bourgeois *self*—and establish and delimit—particularly the deviant, socially pathologized *other*; see Allan Sekula, "The Body and the Archive," *October* 39 (1986): 6–7. Other researchers note a similar context-dependence of the photographic portrait, for instance, the ideological structures of the instrumentalization of photography in dictating its meaning in a concrete political context. The art historian John Tagg is considered another important exponent of this position, because of his Foucauldian- and Marxist-inspired arguments against a medium-specific, essentialist analysis of photography; see John Tagg, *The Burden of Representation: Essays on Photographies and Histories* (Minneapolis: University of Minnesota Press, 1993).

75. A. Sander and A. Döblin, *Antlitz der Zeit: 60 Aufnahmen deutscher Menschen des 20. Jahrhunderts* (München: 1929).

76. Wolfgang Brückle, "Face-Off in Weimar Culture: The Physiognomic Paradigm, Competing Portrait Anthologies, and August Sander's *Face of Our Time*," *Tate Papers*, no. 19 (2013), https://www.tate.org.uk/research/tate-papers/19/face-off-in-weimar-culture-the-physiognomic-paradigm-competing-portrait-anthologies-and-august-sanders-face-of-our-time.

77. Eyal Weizman in "The Bone Cannot Lie," http://www.broombergchanarin.com/article-the-bone.

78. Weizman, http://www.broombergchanarin.com/article-the-bone.

79. Brückle, "Face-Off in Weimar Culture."

80. E.g., B. Jahn, "Deutsche Physiognomik: Sozial-und mediengeschichtliche Überlegungen zur Rolle der Physiognomik in der Weimarer Republik und im Dritten Reich," in *Nach der Sozialgeschichte: Konzepte für eine Literaturwissen-*

schaft zwischen Historischer Anthropologie, Kulturgeschichte und Medientheorie, ed. Martin Huber and Gerhard Lauer (Tübingen, 2000) or M. Uecker, "The Face of the Weimar Republic Photography, Physiognomy, and Propaganda in Weimar Germany," *Monatshefte* 99, no. 4 (2007).

81. Brückle, "Face-Off in Weimar Culture."

82. Brückle.

83. Brückle.

84. P. K. T. Panter, "Auf dem Nachttisch," *Die Weltbühne* 36, no. 466 (1930).

85. The full title of the book was *Menschen der Zeit: Hundert und ein Licht-bildnis wesentlicher Männer und Frauen aus deutscher Gegenwart und jüngster Vergangenheit*, and it was published in Königstein by Karl R. Langewiesche Verlag in 1930. Alfred Döblin wrote the preface, in which, among other things, he emphasized the scientific aspects of Sander's photographic strategy.

86. Brückle, "Face-Off in Weimar Culture."

87. Brückle.

88. Bowker, *Memory Practices in the Sciences,* 183–84.

Chapter 3

1. Karl Marx and Frederick Engels, *Collected Works of Karl Marx and Frederick Engels*, vol. 35 (London: Lawrence & Wishart Electric Books, 2010).

2. Erwin Panofsky, *Tomb Sculpture: Four Lectures on Its Changing Aspects from Ancient Egypt to Bernini*, ed. H. W. Janson and Martin Warnke, ACLS Humanities (New York: H. N. Abrams, 1992), 26–27.

3. Yve-Alain Bois, "On the Uses and Abuses of Look-alikes," *October* 154 (2015): 127.

4. Panofsky, *Tomb Sculpture*, 26–27.

5. However, there is good reason to think that Panofsky may have considered pseudomorphosis and the problem of look-alikes a potentially dangerous part of studying art history. In his article, "On the Uses and Abuses of Look-alikes," Yve-Alain Bois expands this argument, and emphasizes Panofsky's work on detecting *look*-alikes in the Italian renaissance as a sign of his interest in the problem of pseudomorphosis, even though it is never fully developed conceptually.

6. Part of this analysis has previously been published in the Maja Bak Herrie, "Qui est Dominique Lambert?: Sammenligning og virtualitet." In *K&K—Kultur og Klasse*, 47 no. 127 (2019): 71–90.

7. For an elaborated discussion of Stéphanie Solinas's artistic appropriation strategies, see Maja Bak Herrie, "Prodigious Protocols," in *MAST—The Journal of Media Art Study and Theory* 4, no. 2 (2023).

8. E.g., Emily Apter, "Untranslatables: A World System," *New Literary History* 39, no. 3 (2008); Gayatri Chakravorty Spivak, "Rethinking Comparativism," *New Literary History* 40, no. 3 (2009); Rita Felski and Susan Stanford Friedman, *Comparison: Theories, Approaches, Uses* (Baltimore: Johns Hopkins University Press, 2013).

9. E.g., Jay David Bolter and Richard Grusin's seminal book on the aesthetics of remediation, J. David Bolter and Richard A. Grusin, *Remediation: Understanding New Media* (Cambridge, MA: MIT Press, 1999).

10. Gilles Deleuze, *Bergsonism* (New York: Zone, 1988), 96–100.

11. Originally: "Dominique Lambert est un homme de quarante-sept ans. Il est tatillon et méticuleux. Il a des taches de rousseur sur le visage, une peau claire. Ses cheveux sont blancs, mais il était roux étant plus jeune. Il est propre et rasé de près. Son nez est en trompette, légèrement." Artist's translation. See Stéphanie Solinas, *Dominique Lambert* (Paris & Rencontres d'Arles: RVB Books, 2016), n.p.

12. Solinas, *Dominique Lambert*.

13. "Comité Consultatif pour la Description des Dominique Lambert," Solinas, *Dominique Lambert*.

14. Panofsky, *Tomb Sculpture*, 26–27.

15. The distinction between contiguity and similarity is borrowed from Roman Jakobson's linguistic system, which builds on Ferdinand de Saussure's work. Here, contiguity is related to metonymy, whereas similarity is related to the metaphor. See Roman Jakobson, "Two Aspects of Language and Two Types of Aphasic Disturbances," in *Fundamentals of Language*, ed. Moris Halle (Berlin and New York: De Gruyter Mouton, 2002).

16. Ferdinand de Saussure, *Course in General Linguistics*, ed. Charles Bally, Albert Sechehaye, and Albert Reidlinger, trans. Wade Baskin (New York: Philosophical Library, 1959), 123.

17. Solinas, *Dominique Lambert*.

18. Originally: "J'ai défini comme population d'étude les cent quatre-vingt-onze Dominique Lambert répertoriés dans l'annuaire des particuliers (Pages Blanches, France). J'ai élaboré un portait écrit, avec l'aide du Comité Consultatif pour la Description des Dominique Lambert (composé d'un psychologue, un statisticien, un inspecteur de police, un juriste)." Artist's translation. Solinas, *Dominique Lambert, n.p.*

19. Wolfgang Iser, *Die Appellstruktur der Texte, Unbestimmtheit als Wirkungsbedingung literarischer prosa* (Konstanz: Universitatsverlag, 1971).

20. Deleuze, *Bergsonism*, 55.

21. It should be mentioned that in his later work, Deleuze leaves behind the structuralism and semiotics of Saussure, on which my analysis partly draws. Nonetheless, I choose to include Deleuze's concept of virtuality, partly because I believe that it is a great resource for addressing Stéphanie Solinas's artistic research, partly because I believe that it may be argued that parts of Deleuze's thinking are specifically provoked by, but also to some degree continue, the structuralist project. In 1967, Deleuze wrote the essay "How Do We Recognize Structuralism?," which builds on some of the discussions prompted by semiotic thought, see Gilles Deleuze, *Desert Islands and Other Texts, 1953–1974*, ed. David Lapoujade, trans. Michael Taormina (Los Angeles, Cambridge, MA.: Semiotext(e), 2004). Social

and cultural structures do not reside in the individual's mind, nor may they be observed and measured as physical things in the world. Setting aside his late criticism of structuralism, one might argue that Deleuze's work specifically seeks to explains such structures, e.g., Levi R. Bryant, *Difference and Givenness: Deleuze's Transcendental Empiricism and the Ontology of Immanence* (Evanston, IL: Northwestern University Press, 2008), 170–71.

22. Deleuze, *Bergsonism*, 55. The inspiration that Deleuze drew from Alfred North Whitehead's ideas has been explored by only a few scholars (among others, Keith Robinson and Isabelle Stengers). In this section, in particular, one notices this inspiration, as Deleuze draws on what Robinson describes as Whitehead's "ontological constructivism" or "speculative empiricism" (that is, a post-Kantian understanding of experience, beginning with the objects of real experience followed by a "deduction" of their genetic processes and ontological conditions); see Keith Robinson, "The New Whitehead?," *Symposium: Canadian Journal of Continental Philosophy* 10, no. 1 (2006): 71–72.

23. Deleuze, *Bergsonism*, 96.

24. David Hills, "Metaphor," in *The Stanford Encyclopedia of Philosophy*, ed. Edward N. Zalta and Uri Nodelman (Metaphysics Research Lab: Stanford University, 2022), https://plato.stanford.edu/archives/fall2024/entries/metaphor/.

25. This interest in patterns and connections in data is far from new, just as the challenges of explaining and legitimizing the existence of patterns met in nature, history, mathematics, or in an individual authorship are well known. However, computational pattern recognition does not have many years behind it, and it is this mode of comparison I am interested in here. Chiel van den Akker elaborates on the difference between pattern recognition in the traditional sense in the humanities and "computer-aided" pattern recognition; see Chiel van den Akker, "What Are Patterns in the Humanities?," *Interdisciplinary Science Reviews* 43, no. 1 (2018).

26. Kathleen Fitzpatrick, "The Humanities, Done Digitally," in *Debates in the Digital Humanities*, ed. Matthew K. Gold (Minneapolis: University of Minnesota Press, 2012).

27. Matthew G. Kirschenbaum, "What Is Digital Humanities and What's It Doing in English Departments?," *ADE Bulletin* 150 (2010).

28. Mark B. N. Hansen, "Our Predictive Condition, or, Prediction in the Wild," in *The Non-Human Turn*, ed. Richard Grusin (Minneapolis: University of Minnesota Press, 2015), 108.

29. For a discussion of the nature of patterns and their status as an epistemological object in the field of digital humanities, see Dan Dixon, "Analysis Tool or Research Methodology: Is There an Epistemology for Patterns?" *Understanding Digital Humanities* (London, 2012), 191–209.

30. Notable works that generally testify to such a defensive position on the deployment of sophisticated machine techniques include Franco Moretti, *Distant*

Reading (London: Verso, 2013); Matthew L. Jockers, *Macroanalysis: Digital Methods and Literary History* (Urbana, IL: University of Illinois Press, 2013); Matthew Wilkens's article "The Geographic Imagination of Civil War-Era American Fiction," *American Literary History* 25 (Winter 2013): 803–40; and Andrew Piper and Mark Algee-Hewitt's article "The Werther Effect I: Goethe, Objecthoods, and the Handling of Knowledge," in *Distant Readings: Topologies of German Culture in the Long Nineteenth Century*, ed. Matt Erlin and Lynn Tatlock (Rochester, NY: Camden House, 2014), 155–84.

31. van den Akker, "What Are Patterns in the Humanities?," 83.

32. Rob Kitchin, "Big Data, New Epistemologies and Paradigm Shifts," *Big Data & Society* 1, no. 1 (2014): 4.

33. Kitchin is especially critical of Chris Anderson's problematic (and at this point, infamous and often-cited) essay "The End of Theory," in which he writes: "There is now a better way. Petabytes allow us to say: 'Correlation is enough.' . . . We can analyze the data without hypotheses about what it might show. We can throw the numbers into the biggest computing clusters the world has ever seen and let statistical algorithms find patterns where science cannot . . . Correlation supersedes causation, and science can advance even without coherent models, unified theories, or really any mechanistic explanation at all. There's no reason to cling to our old ways." See Chris Anderson, "The End of Theory: The Data Deluge Makes the Scientific Method Obsolete," *Wired*, June 23, 2008, https://www.wired.com/2008/06/pb-theory/. Other voices are also included, among others, Jill Dyche, who argues that when you "mine" Big Data, you may uncover "relationships and patterns that we didn't even know to look for." See Jill Dyche, "Big Data 'Eurekas!' Don't Just Happen," *Harvard Business Review*, November 20, 2012, https://hbr.org/2012/11/eureka-doesnt-just-happen. Also, one might consider Ian Steadman's ideas that "[n]othing is lost from looking too closely at one particular section of data; nothing is lost from trying to get too wide a perspective on a situation that the fine detail is lost," and that "[t]he analyst doesn't even have to bother proposing a hypothesis anymore." See Ian Steadman, "Big Data and the Death of the Theorist," *Wired*, January 25, 2013.

34. This construction is found in the areas of the natural sciences and computer sciences that are occupied with so-called data-driven or data-intensive research. This construction and the changes in methodological approach that it represents differ from more traditional, experimental, or deductive approaches, where hypotheses and insights are developed against a background of theory, rather than on "knowledge" from a data set. E.g., Steve Kelling et al., "Data-Intensive Science: A New Paradigm for Biodiversity Studies," *BioScience* 59, no. 7 (2009), https://doi.org/10.1525/bio.2009.59.7.12; Kitchin, "Big Data, New Epistemologies and Paradigm Shifts," 5–6. However, it is important to emphasize that Kitchin distinguishes between data-driven science and more inductive processes, because to him, data-driven science is a marginal figure in the field, which also includes

deductive (and perhaps even abductive) approaches. This argument is developed in Simon Aagaard Enni and Maja Bak Herrie's, "Turning Biases into Hypotheses Through Method: A Logic of Scientific Discovery for Machine Learning," *Big Data & Society* 8, no. 1 (2021).

35. E.g., Rob Kitchin, "Big Data and Human Geography: Opportunities, Challenges and Risks," *Dialogues in Human Geography* 3, no. 3 (2013); Kitchin, "Big Data, New Epistemologies and Paradigm Shifts"; Kate Crawford, "The Hidden Biases in Big Data," *Harvard Business Review*, April 1, 2013.

36. Franco Moretti, *Graphs, Maps, Trees: Abstract Models for a Literary History* (London: Verso, 2007), 53 (Moretti's emphases).

37. Andrew Piper, *Enumerations: Data and Literary Studies* (Chicago: University of Chicago Press, 2020): 7.

38. Piper, *Enumerations*, xi.

39. Moretti, *Distant Reading*, 66–67. The phase "great unread" is attributed to Margaret Cohen, who coined it.

40. For critiques by scholars such as Alexander Galloway, David Golumbia, Tara Mcpherson, all from the field of new media studies, see the special issue of *Differences* on the theme of "In the Shadows of the Digital Humanities" (*Differences* 25, no. 1 (2014). Their essays include Alexander Galloway, "The Cybernetic Hypothesis," pp. 107–31; David Golumbia, "Death of a Discipline," pp. 156–76; and Tara McPherson, "Designing for Difference," pp. 177–88.

41. Daniel Allington, Sarah Brouillette, and David Golumbia, "Neoliberal Tools (and Archives): A Political History of Digital Humanities," *Los Angeles Review of Book*, May 1, 2016.

42. Ann Blair, *Too Much to Know: Managing Scholarly Information Before the Modern Age* (New Haven, CT: Yale University Press, 2011).

43. David M. Berry, "The Computational Turn: Thinking About the Digital Humanities," *Culture Machine* 12 (2011): 2.

44. Hills, "Metaphor," https://plato.stanford.edu/archives/fall2024/entries/metaphor/.

45. danah boyd and Kate Crawford, "Six Provocations for Big Data," A Decade in Internet Time: Symposium on the Dynamics of the Internet and Society, September 21, 2011.

46. Emily Apter, "Translating Extraction: Archaeologies of Knowledge, Data-Mining, Overburden," (Situated Knowing: The Economies of Representations, New York City, May 10, 2018).

47. Aaron L. Mishara, "Klaus Conrad (1905–1961): Delusional Mood, Psychosis, and Beginning Schizophrenia," *Schizophrenia Bulletin* 36, no. 1 (2010): 11.

48. Charles Sanders Peirce, *The Collected Papers of Charles Sanders Peirce*, electronic ed., ed. Charles Hartshorne, Paul Weiss, and Arthur W. Burks, vol. 8, *Reviews, Correspondence, and Bibliography* (Cambridge, MA: Belknap Press of Harvard University Press, 1994), 8.342.

49. Peirce, *The Collected Papers of Charles Sanders Peirce*, 8.354.

50. Peirce, 8.328.

51. Wim Staat, "On Abduction, Deduction, Induction and the Categories," *Transactions of the Charles S. Peirce Society* 29, no. 2 (1993): 28.

52. Peirce, *The Collected Papers of Charles Sanders Peirce*, 5.157.

53. Jeffrey R. DiLeo, "Peirce's Haecceitism," *Transactions of the Charles S. Peirce Society* 27, no. 1 (1991), 89.

54. See Kitchin, "Big Data, New Epistemologies and Paradigm Shifts," 4. Another great source is Luciana Parisi's Peirce-inspired critique of induction in machine learning, Luciana Parisi, "Critical Computation: Digital Automata and General Artificial Thinking," *Theory, Culture & Society* 36, no. 2 (2019). This critique includes references to Pedro Domingo's seminal book, *The Master Algorithm* (2015), and Mareille Hildebrandt and Antoinette Rouvroy's work on autonomic computing; see Mareille Hildebrandt and Antoinette Rouvroy, *Law, Human Agency and Autonomic Computing: The Philosophy of Law Meets the Philosophy of Technology* (Milton Park, Abingdon, Oxon: Routledge, 2011). Both texts are critical of induction in computation today, see p. 7 and 126 respectively.

55. Or, according to Peirce's vocabulary, the form of *Copulants* as they relate to *Descriptives* and *Designatives*.

56. For example, see the extensive work done by the *Uncertain Archives* research group at the University of Copenhagen (2015–2019). A great place to start might be with the vast, interdisciplinary exploration of big data archival practices in Nanna Bonde Thylstrup et al.'s, *Uncertain Archives: Critical Keywords for Big Data* (Cambridge, MA: MIT Press, 2021). Other important voices include (among many others), Luciano Floridi, "Big Data and Their Epistemological Challenge," *Philosophy & Technology* 25, no. 4 (2012); David Bollier and Charles M. Firestone, *The Promise and Peril of Big Data* (Washington, DC: Aspen Institute, Communications and Society Program, 2010); Viktor Mayer-Schönberger and Kenneth Cukier, *Big Data: A Revolution That Will Transform How We Live, Work, and Think* (Boston: Mariner Books, 2014).

57. This argument is elaborated in Enni and Herrie's "Turning Biases into Hypotheses through Method: A Logic of Scientific Discovery for Machine Learning."

58. Kitchin, "Big Data, New Epistemologies and Paradigm Shifts," 5.

59. Peirce, *The Collected Papers of Charles Sanders Peirce*, 5, 1.328.

60. As Lorraine Daston (here, via Foucault, and not Peirce) states: "Indigestible first nature becomes intelligible second nature, and the scientific work of hypothesizing, testing, explaining, and predicting can begin. But once second nature slips from science present into science past, collective empiricism requires third nature: the repository of those findings of second nature selected to endure. These are the archives of the sciences." See Lorraine Daston, "Third Nature," in *Science in the Archives: Pasts, Presents, Futures*, ed. Lorraine Daston (Chicago: University of Chicago Press, 2017), 1.

61. As she writes in the introduction, "Precisely ten years after receiving the passport photographs from the Dominique Lamberts, the possible faces of the 'true' Dominique Lamberts are disclosed, offered to the gaze in *Dominique Lambert*." See Solinas, *Dominique Lambert*.

62. Moretti, *Graphs, Maps, Trees*, 1–2.

63. Lisa Gitelman, *"Raw Data" Is an Oxymoron*, Infrastructures Series (Cambridge, MA: MIT Press, 2013), 1.

64. Daston, "Third Nature," 1.

Conclusion

1. K. Green, "Deixis and Anaphora: Pragmatic Approaches," in Encyclopedia of Language and Linguistics, 2nd ed., ed. Keith Brown, 415–17 (London: Elsevier, 2006).

2. See Karl Marx and Frederick Engels, *Collected Works of Karl Marx and Frederick Engels*, vol. 28 (London: Lawrence & Wishart Electric Books, 2010), 49.

3. Green, "Deixis and Anaphora."

4. Liam Cole Young, "Cultural Techniques and Logistical Media: Tuning German and Anglo-American Media Studies," *M/C journal* 18, no. 2 (2015); Cornelia Vismann, *Files: Law and Media Technology*, trans. Geoffrey Winthrop-Young, Meridian (Stanford, CA: Stanford University Press, 2008), 5–6.

5. Daniel Rosenberg, "Stop, Words," *Representations* 127, no. 1 (2014).

6. Sybille Krämer, "The Cultural Techniques of Time Axis Manipulation: On Friedrich Kittler's Conception of Media," *Theory, Culture & Society* 23, no. 7/8 (2006).

7. Geoffrey C. Bowker and Susan Leigh Star, *Sorting Things Out: Classification and Its Consequences*, 4th ed., Inside Technology (Cambridge, MA: MIT Press, 2002), 11.

8. Donna Haraway, "Situated Knowledges: The Science Question in Feminism and the Privilege of Partial Perspective," *Feminist Studies* 14, no. 3 (1988).

9. Giorgio Agamben, *What Is an Apparatus? And Other Essays* (Stanford, CA: Stanford University Press, 2009),, 2.

10. Jørn Erslev Andersen, *Sansning og erkendelse: Æstetikhistoriske grundtekster fra Baumgarten til Kant* (Aarhus: Aarhus Universitetsforlag, 2012), 226.

11. A. S. Aurora Hoel, "Lines of Sight: Peirce on Diagrammatic Abstraction," in *Das Bildnerische Denken: Charles S. Peirce*, ed. Franz Engel, Moritz Queisner, and Tullio Viola (Berlin/Boston: De Gruyter, 2012).

12. To many, the distinction between subject and object, or between the subjective and the objective, has become synonymous with the distinction between consciousness and the material world. However, this understanding was not always the most common. For example, subject, which for Descartes signified self-consciousness as an epistemological principle, originally meant the same as substance (in Latin, *substantia*), which was a translation of the Greek word *hypo-*

keimenon. In this older usage of language, *subjectum* corresponds to a "substrate" or a "substance" that carries something. The opposite of *subjectum* is *objectum*, which, instead of denoting "actual," readily available "things," describes something imagined or thought. Although I do not consistently use this older definition of the object in this book, but instead incline to the theoretical object as a concept, I believe it is evident here that the term *object* does not necessarily imply something physical or thing-like, as opposed to a more dynamic consciousness.

13. Hoel, "Lines of Sight," 254.

14. Hoel, "Lines of Sight," 271.

15. Charles S. Peirce, *The New Elements of Mathematics: Algebra and Geometry* (Germany: De Gruyter, 2016), 324. Cited in Hoel, "Lines of Sight," 271.

INDEX

abduction: abductive approach, 9, 101–102; hypothesis, 10, 11–14, 44–46, 80, 95; hypothesizing, 100–101; theory of abduction 11–12

abstraction: to abstract; 10, 97, 105–106, 109; abstraction in art, 25–26; conceptual abstraction, 7, 13, 17, 45, 58, 101–103, 108, 110, 116; practices of abstracting, 5, 13–14, 18, 37, 70, 99, 106; statistical abstraction, 3–4

aesthetics: aestheticization, 36, 75; aesthetic attitude, 3, 16, 122n9; aesthetic character, 36; aesthetics as sense-perception (*aisthêsis*) 3, 26, 116, 122n8, 131n30; aesthetics of concepts, 35–36; aesthetic discipline, 15, 108–109, 115–119, 122n8, 124n34, 131n30; aesthetic knowledge production 113, 117, 119, 122n6; aesthetic manifestation, 4, 20–25, 35–38, 44, 74, 77–78, 104; aesthetic mediation, 111–113; aesthetic operations, 4, 13, 17, 22, 109; aesthetic phenomenon, 3, 9; aesthetic process 46; aesthetic relations, 11, 15, 37, 59, 119

Agamben, Giorgio, 23, 123n13

aggregate: aggregation modes, 46–48, 51, 56–71, 75–76, 78; aggregating practices, 42–43, 54–71, 75–76, 87–88, 96, 114; aggregation space, 40; aggregates as statistical entity, 1–2, 5, 16–17, 77–78, 108, 115

apparatus. *See* dispositif

algorithm, 22, 107–108, 132n43; classification algorithm, 22, 31; algorithmic processing, 34–36, 56, 65, 69, 106, 119; algorithmic thought, 23, 44, 78, 112

Antlitz der Zeit. See Sander, August

apophenia, 100, 105

archive, 71, 75–76; practices of arching, 7, 33, 71

art, 3, 16; artistic practice 4, 15–16, 23, 82, 91, 96, 116; artistic research 16, 85, 126n53

••• **Sensing Media**
Aesthetics, Philosophy,
and Cultures of Media
EDITED BY WENDY HUI KYONG CHUN
AND SHANE DENSON

What does it mean to think, feel, and sense with and through
media? In this cross-disciplinary series we present books and
authors exploring this and related questions: How do media
technologies, broadly defined, transform artistic practices and
aesthetic sensibilities? How are practices, encounters, and affects
entangled with the deep infrastructures and visible surfaces of
the media environment? How do we "make sense"—cognitively,
perceptually, and culturally—of media?

We are especially interested in contributions that open our
understanding of media aesthetics beyond the narrow confines of
Western art and aesthetic values. We seek works that reestablish the
environmental connections between art and technology as well as
between the aesthetic, the sensible, and the philosophical. We invite
alternative epistemologies and phenomenologies of media rooted in
the practices and subjectivities of Black, Indigenous, queer, trans,
and other communities that have been unjustly marginalized in
these discussions. Ultimately, we aim to sense the many possible
worlds that media disclose.

—

The authorized representative in the EU for product safety and compliance is:
Mare Nostrum Group B.V.
Mauritskade 21D
1091 GC Amsterdam
The Netherlands
Email address: gpsr@mare-nostrum.co.uk

KVK chamber of commerce number: 96249943

The authorized representative in the EU for product safety and compliance is:
Mare Nostrum Group
B.V Doelen 72
4831 GR Breda
The Netherlands

www.ingramcontent.com/pod-product-compliance
Lightning Source LLC
Chambersburg PA
CBHW050448290526
45786CB00006B/2211